Caribou

Common
shrew

Raccoon

Snowshoe hare

Virginia opossum

Tiger

National Geographic

BOOK OF
Mammals
Volume One

Prepared by the
Special Publications
Division

National
Geographic Society,
Washington, D. C.

NATIONAL GEOGRAPHIC BOOK OF MAMMALS

Published by
  The National Geographic Society
  Gilbert M. Grosvenor, *President*
  Melvin M. Payne, *Chairman of the Board*
  Owen R. Anderson, *Executive Vice President*
  Robert L. Breeden, *Vice President, Publications
    and Educational Media*

Prepared by
  The Special Publications Division
  Donald J. Crump, *Editor*
  Philip B. Silcott, *Associate Editor*
  William L. Allen, William R. Gray, *Senior Editors*

  Joan Tapper, *Managing Editor*
  Karen Altpeter Zogg, *Picture Editor*
  Suez B. Kehl, *Art Director*
  Cinda Rose, *Assistant Art Director*
  Michael Barrier, Jane R. McGoldrick, *Editor-Writers*
  Suzanne de Lesseps, Joyce Diamanti, Elizabeth W. Fisher,
    Jacqueline Geschickter, Karen M. Kostyal, Mary Ann
    Larkin, Carolyn McKee, Grace E. Moremen, Lisa Olson,
    Catherine O'Neill, Laurel H. Rabin, Donna Turner, Peggy
    D. Winston, Suzanne Venino, *Writers*
  Karen M. Kostyal, *Senior Researcher and Assistant to the Editor*
  Debra A. Antonini, Sarah Clark, Margo Crabtree,
    Elizabeth W. Fisher, Palmer Graham, Catherine
    Herbert-Howell, Carolinda Hill, Stephen Hubbard,
    Jean Kaplan-Teichroew, Lucie Langford, Patricia Y.
    Larkin, Carolyn McKee, Peggy D. Winston, *Researchers*

Illustrations and Design
  Jody Bolt, *Consulting Art Director*
  Turner Houston, Marianne R. Koszorus, *Designers;*
    Holly Bowen, Richard Fletcher, Cynthia B. Scudder,
    Design Assistants
  Ned Smith, *Black-and-White Illustrations*
  Darrell K. Sweet, *Paintings*
  John D. Garst, Jr., Peter J. Balch, Lisa Biganzoli, Patricia K.
    Cantlay, Margaret Deane-Gray, Gary M. Johnson, Susan
    M. Johnston, Alfred L. Zebarth, *Map Research, Design,
    and Production*

Engraving, Printing, and Product Manufacture
  Robert W. Messer, *Manager*
  George V. White, *Production Manager*
  Raja D. Murshed, *Production Project Manager*
  Mark R. Dunlevy, Richard A. McClure, Christine A.
    Roberts, David V. Showers, Gregory Storer, *Assistant
    Production Managers*
  Susan M. Oehler, *Production Staff Assistant*
  Debra A. Antonini, Nancy F. Berry, Bonnie L. Biddix,
    Pamela A. Black, Barbara Bricks, Jane H. Buxton,
    Mary Elizabeth Davis, Eve E. Galloway, Rosamund
    Garner, Victoria D. Garrett, Nancy J. Harvey, Suzanne
    J. Jacobson, Lori Kehl, Artemis S. Lampathakis,
    Sandra Lee Matthews, Virginia A. McCoy, Amy E.
    Metcalfe, Merrick P. Murdock, Cleo Petroff, Linda J.
    Rinkinen, Marcia Robinson, Carol A. Rocheleau,
    Maria A. Sedillo, Katheryn M. Slocum, Musette Steck,
    Jenny Takacs, Phyllis C. Watt, *Staff Assistants*

  Anne K. McCain, *Index*

  Glenn O. Blough, Professor Emeritus of Science Education
    University of Maryland; Judith Hobart, Librarian,
    Beauvoir, The National Cathedral Elementary School;
    Henry W. Setzer, Curator of Mammals, Emeritus,
    Smithsonian Institution, *Consultants*

Oribi in southeastern Africa quietly feeds among tall grasses. If startled, this small member of the antelope family will bound swiftly away.

PRECEDING PAGES: *Heads up! Here comes an elephant—largest of all land mammals. Behind it, a parade of adults and young crosses a grassy plain in Africa.* PAGE 1: *Young mountain goat gallops headlong down a rocky slope in Washington State.* ENDPAPERS: *Making tracks, mammals big and small leave their footprints.*
COVER: *Young chimpanzee in Tanzania, a country in Africa, balances between two vines.*

Foreword

FOR ME—and I suspect for almost everyone—mammals appeal deeply to the imagination. One of my favorites is the elephant. It's hard to imagine that such a massive beast can manipulate its trunk to pick up a tiny peanut and yet be able to heft a thousand-pound log of teak with obvious ease! In Sri Lanka, Raja, a movie-star elephant, once swung me up with his trunk and carried me across a river. That night, Raja gently led a sacred procession through a jammed crowd of ten thousand people—without stepping on anyone's feet.

Many people—my daughter, Alexi, among them—are intrigued by cats of any size. Others favor dogs, and still others are enthralled by whales and other marine mammals. This interest is natural, for we too are mammals—tied to the other members of this realm by myriad, complex relationships.

For those of us privileged to observe animals in the wild and to interact with them, the allure of the world of mammals is even greater. While snorkeling with humpback whales off Hawaii, I was amazed by these gentle giants—among the largest mammals ever to inhabit earth. I watched one huge humpback swim—carefully it seemed—over a diver in its path. The mere slap of a fluke could have instantly killed the diver. As the whales glided silently past, I marveled at their sensitivity toward other creatures.

In Africa, I have followed herds of wildebeests across the Serengeti Plain. How do these large, shaggy antelopes live, forage for food, escape predators? Why do they move along established migratory paths, when other kinds of antelopes do not? How many kinds of antelopes are there? And what are their differences and similarities?

These are the kinds of questions that the two-volume *Book of Mammals* answers. Educators, librarians, and specialists in many fields of mammalogy have contributed their expertise. In-depth information about habitat, food, and behavior—presented in a form that the entire family can enjoy—makes these volumes unique. They greatly expand the opportunities for parents like me to teach our children about the world of mammals.

Illustrated with nearly a thousand exciting photographs of animals in the wild, the *Book of Mammals* is much more than a collection of wildlife portraits and a compilation of vital statistics. It is an extraordinary work, compelling in its breadth of information, in its variety, and in the quality and the range of its photographs.

Never in my 25 years with the National Geographic Society have I been so enthusiastic about a project. Since the Geographic's founding nearly a hundred years ago, Society members have ranked wild animals and animal behavior among their chief interests. I believe that the *National Geographic Book of Mammals* sets a standard of excellence in the field of wildlife study. It not only reveals the heritage of the wild, free past, but it also points the way to greater understanding and concern for a new generation of readers.

GILBERT M. GROSVENOR
President
National Geographic Society

Contents

Coats damp from fishing in a river in Alaska, a female brown bear and her cub pause on a gravel bar.

Volume One

Volume Two

How To Use the
Book of Mammals

IN THE *National Geographic Book of Mammals,* you will learn about the habits and behavior of mammals of the world. Each entry concentrates on a single kind of animal such as the aardvark or on a closely related group of mammals such as monkeys. To learn about mammals in general, read "What Is a Mammal?" beginning on page 10. This in-depth essay, written by respected mammalogist Henry W. Setzer, explains why scientists classify certain animals as mammals. It compares some of the ways that mammals find food and shelter, raise their young, defend themselves, and communicate.

The pages of the *Book of Mammals* are numbered consecutively from page 1 in Volume One to page 608 in Volume Two. A glossary appears in Volume Two on page 600. It defines words that may be unfamiliar to some readers. For additional information on a number of specific mammals or about mammals in general, refer to the list of additional reading on page 603. On page 604, the entries in both volumes are grouped by the scientific order to which the mammals belong. The chart on that page also includes information on what distinguishes one order from another. An index begins on page 605. It lists the common and scientific names of species of mammals discussed or pictured in the *Book of Mammals.*

On page 9, a sample entry for the ibex appears. It is slightly different in appearance from the ibex entry on page 286. It has been changed to bring together in one place all the elements that might occur on several pages. The numbered paragraphs below match the numbered elements in the sample entry. These elements appear consistently throughout the *Book of Mammals.*

(1) The *Book of Mammals* is organized alphabetically. Each alphabetical section begins with a large letter at the top of a left-hand page.

(2) Each entry is headed by the common name of the mammal covered on that page. Some entries, like the one on the cheetah, cover only a single species. Others, like the one on the ibex, include several species. Still others, like the entry on the llama, cover several related mammals—llamas, alpacas, guanacos, and vicuñas.

(3) With the heading of the entry, you'll find a pronunciation guide for the animal's name. For ease in pronunciation, the guides use common words and syllables. The syllable in capital letters should be accented. For simple names like "goat" or "bear" there is no pronunciation guide. Difficult words in the entries also are followed by pronunciation guides.

(4) Every entry includes a range map. The colored areas on the maps show where the animal lives throughout the world. Some animals live in many places. Others have limited ranges. The maps do not include the ranges of domestic animals—pets or those that are kept on farms.

Each of the 19 orders, or groups, of mammals has a different color for its range maps. The range of every mammal in that order is shown in that color. For example, the ibex is an artiodactyl (say art-ee-oh-DAK-tul), or even-toed hoofed mammal. Its range is shown in green. Similarly, the ranges of all the artiodactyls appear in the same shade of green. The colors used for the orders are shown on page 17.

(5) Each entry in the *Book of Mammals* includes a fact box that provides information about all the species of the animals in the entry. "Height" or "Length of Head and Body" indicates the size of the animals. "Weight" tells how heavy they can be.

"Habitat and Range" describes the terrain the animals live in and the parts of the world they inhabit. "Food" lists what the animals eat. "Life Span" reveals how long the animals live. "Reproduction" tells how long a pregnancy lasts for the animals and how many young are born at one time. "Order" shows the scientific classification of the animals.

6 All sizes and distances in the *Book of Mammals* are given both in standard United States measurements and in metric measurements. All metric conversions are rounded off to whole numbers.

7 Every photograph has a caption that describes the picture. For clarity, captions often are keyed to photographs by triangles.

8 Identification lines give the common names of the animals shown in photographs and the average sizes of adult animals. This information is given only once. If the entry includes only one species, there is no identification line. See the fact box for this information.

9 Cross-references are given when the information about an animal appears under another name or with the entry for another animal. Check the index for other animals that interest you.

What Is a Mammal?

By Henry W. Setzer
Curator of Mammals, Emeritus, Smithsonian Institution

FROM DOGS AND CATS to elephants and kangaroos—the world of mammals is filled with incredible diversity. Graceful porpoises, which spend their lives in the water, are mammals. So are bats, which fly through the night air. Tall giraffes and tiny mice are mammals, as are chattering monkeys and powerful tigers. Human beings are mammals, too.

What characteristics do all these animals share? What sets them apart from the other creatures that belong to the animal kingdom?

Scientists break the animal kingdom into two divisions: vertebrates (say VURT-uh-bruts), or animals with backbones, and invertebrates (say IN-vurt-uh-bruts), those without backbones. Mammals are vertebrates, but so are fishes, amphibians, reptiles, and birds. Of these animals, however, only birds and mammals are warm-blooded. That means their bodies stay at almost the same temperature even when temperatures around them vary widely.

But birds and mammals have different kinds of body coverings. Birds have feathers, and mammals have hair. Of all the animals in the world, only mammals have hair, though not all hair looks the same. Whales have only a few coarse hairs near their mouths. Thick, curly wool covers the bodies of domestic sheep. Many pigs have scattered, coarse bristles. Hedgehogs have spines, and porcupines have quills. Fine fur covers cats and foxes.

Other characteristics besides hair set mammals apart from all other animals. An important trait is that every female mammal feeds her young on milk from her body. The word mammal comes from the Latin word *mamma*, which means "breast."

Scientists think that mammals first appeared

Nose-deep in grass, a lion watches for prey. Scientists ▷ group lions among the carnivores, or meat eaters.

What Is a Mammal?

*Prehensile-tailed porcupine, found in Central and ▷
South America, sniffs a branch as it climbs. This tree-
dweller has quills mixed with its other hair.
▽ Hanging upside down, a three-toed sloth of South
America moves slowly but easily through the trees.*

△ *Australian hopping mouse licks its delicate paws.
When startled, this rodent bounds away on its hind legs.*

more than 200 million years ago. They descended from a group of reptiles called therapsids (say thuh-RAP-suds). Therapsids resembled mammals in the number of bones in their feet and in the way the jaw muscles were attached to their skulls. A few therapsids may have been warm-blooded, had hair, and fed their young on milk. Scientists do not know for sure. Descendants of reptiles that had these characteristics, among others, became the first mammals.

Today there are thousands of species, or kinds, of mammals. Scientists have put them into 19 groups called orders. Which order a mammal belongs to may depend on the arrangement of its bones or the kind of teeth it has. It may also depend on how the mammal bears its young.

Monotremes (say MON-uh-treemz) share several characteristics with reptiles. They lay leathery-shelled eggs that hatch into young animals—instead of bearing live young. Like all mammals, however, monotreme young are nourished by their mothers' milk. Instead of sucking from a nipple, a young monotreme gets milk from pores on its mother's belly. The only members of the monotreme order are echidnas and platypuses.

Marsupials (say mar-soo-pea-ulz) give birth to living young that are tiny and underdeveloped. At birth, most marsupials crawl into a pouch or a sac on the underside of their mothers' bodies. Inside, each attaches itself to a nipple. There it remains, drinking milk and growing stronger. When it is more fully developed, it begins to spend time out of the pouch.

Many different kinds of marsupials live in Australia. Koalas and kangaroos are probably the most familiar. In the Western Hemisphere, only opossums belong in this order of mammals.

Mammals in all the other orders give birth to living young that can survive outside their mothers' bodies as soon as they are born. The offspring get their nourishment by nursing.

*Curling back her lips, a huge female black rhinoceros ▷
stands ready to defend her calf.*

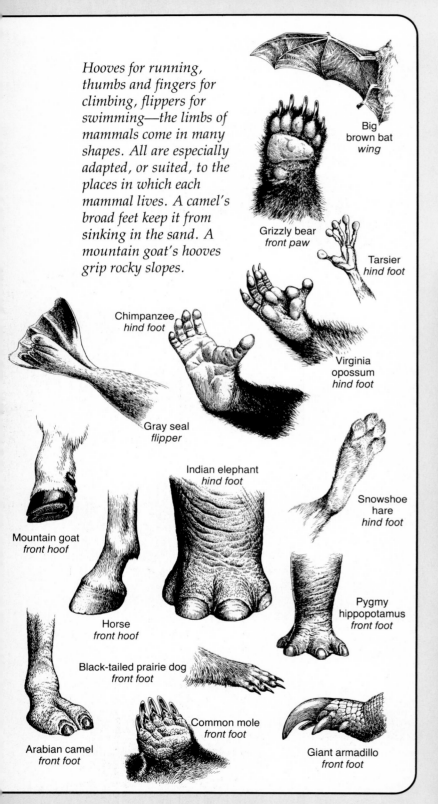

Hooves for running, thumbs and fingers for climbing, flippers for swimming—the limbs of mammals come in many shapes. All are especially adapted, or suited, to the places in which each mammal lives. A camel's broad feet keep it from sinking in the sand. A mountain goat's hooves grip rocky slopes.

Big brown bat
wing

Grizzly bear
front paw

Tarsier
hind foot

Chimpanzee
hind foot

Virginia opossum
hind foot

Gray seal
flipper

Indian elephant
hind foot

Snowshoe hare
hind foot

Mountain goat
front hoof

Horse
front hoof

Pygmy hippopotamus
front foot

Black-tailed prairie dog
front foot

Arabian camel
front foot

Common mole
front foot

Giant armadillo
front foot

14

Some orders include hundreds of species—chiropterans (say kye-ROP-tuh-runs), or bats, for example. Others—like tubulidentates (say too-byu-luh-DEN-tates), or aardvarks—include only one.

Sometimes it is easy to see why animals are grouped together. Carnivores (say CAR-nuh-vorz) are basically meat eaters. Insectivores (say in-SEK-tuh-vorz) eat mainly insects. Artiodactyls (say art-ee-oh-DAK-tulz) are hoofed animals with an even number of toes on each foot. The illustration on page 17 lists the orders of mammals and shows one representative of each order. A chart on page 604 tells

▽ *Bulky manatee swims in a river in Florida. These mammals spend all their lives in the water.*

Wings partly folded, a yellow-eared tent bat roosts on ▷ a branch in Panama. Of all mammals, only bats can fly.

what makes each order special. It lists all the entries in these volumes by order.

Mammals live in almost every climate and terrain. Because their bodies stay about the same temperature in widely varying conditions, mammals do not rely on the sun to keep warm. They can move about after dark and in the cold, when cold-blooded animals seek shelter. The hair that covers most mammals serves as a blanket against the cold. When they get too warm, many mammals sweat or pant. This helps lower their body temperatures.

Armored with scales, a Cape pangolin (above, left) crosses a dry grassland in Africa. A brush-tailed possum (above, right) pauses in a tree fork in Australia. Like most marsupials, this possum has a pouch for its young.

Mammals are adapted, or suited, to their environments in many other ways. Polar bears and some seals have layers of fat that protect them in the icy climate of the polar regions. The streamlined bodies of whales and porpoises move easily through the ocean. Though they must come to the surface to breathe, these marine mammals spend all their lives in the water. Animals that live in deserts usually need little water. Kangaroo rats and fennecs get most of the moisture they need from their food. High on mountain slopes, the tiny pika survives by drying plants in summer to eat when food is scarce.

The feet, hooves, or paws of a mammal may give a clue to where the animal lives. Camels have broad, padded feet that keep them from sinking into sand. The padded hooves of mountain goats give them a good grip on steep, rocky slopes. An armadillo's thick front claws help the animal dig through

15

Fine yellowish hair frames the dark face of a golden ▷ langur, a kind of monkey.

sunbaked earth. A tarsier's padded fingertips help the tiny primate cling to tree trunks.

Sometimes animals become extinct, or die out, because they no longer fit into their environment. When the climate or the terrain changes sharply over decades or even centuries, the animals may not be able to adapt and to find the food and shelter they need. Millions of years ago, the hornless rhino probably died out when the trees it ate gradually disappeared. Because of the way its body had developed, it could not eat any other kind of food. The woolly mammoth too may have disappeared because of changes in climate at the end of the Ice Age.

People around the world have developed some special breeds of mammals. Domestic, or tame, animals—cows, horses, dogs, cats, goats, and sheep—all have been bred to provide things people want or need. Cows can be bred to produce increased amounts of milk or meat. Sheep can be bred for thicker coats. Cats and dogs can be bred simply for traits that please their owners.

On the following pages you can read more about how mammals in the wild survive in different environments. Though they all eat, move, find shelter, and care for young, mammals do these things in an amazing variety of ways.

Caribou trots through low shrubs. Like most male deer, caribou grow bony antlers every year.

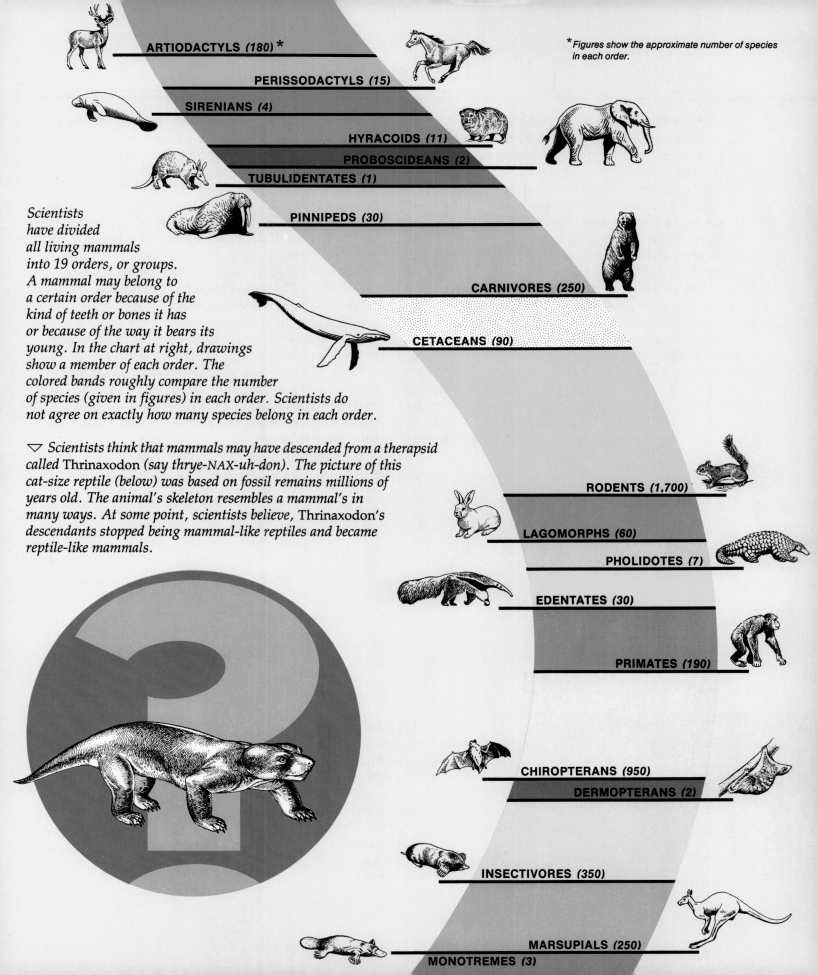

ARTIODACTYLS *(180)* *

PERISSODACTYLS *(15)*

SIRENIANS *(4)*

HYRACOIDS *(11)*

PROBOSCIDEANS *(2)*

TUBULIDENTATES *(1)*

PINNIPEDS *(30)*

CARNIVORES *(250)*

CETACEANS *(90)*

RODENTS *(1,700)*

LAGOMORPHS *(60)*

PHOLIDOTES *(7)*

EDENTATES *(30)*

PRIMATES *(190)*

CHIROPTERANS *(950)*

DERMOPTERANS *(2)*

INSECTIVORES *(350)*

MARSUPIALS *(250)*

MONOTREMES *(3)*

* *Figures show the approximate number of species in each order.*

Scientists have divided all living mammals into 19 orders, or groups. A mammal may belong to a certain order because of the kind of teeth or bones it has or because of the way it bears its young. In the chart at right, drawings show a member of each order. The colored bands roughly compare the number of species (given in figures) in each order. Scientists do not agree on exactly how many species belong in each order.

▽ Scientists think that mammals may have descended from a therapsid called Thrinaxodon (say thrye-NAX-uh-don). The picture of this cat-size reptile (below) was based on fossil remains millions of years old. The animal's skeleton resembles a mammal's in many ways. At some point, scientists believe, Thrinaxodon's descendants stopped being mammal-like reptiles and became reptile-like mammals.

Homes and Habitats

IN FREEZING POLAR REGIONS or in steamy tropics—wherever they live, mammals need places to take shelter from snow and ice, from sun and wind. They also need places to hide from enemies, to raise young, to store food, and to sleep.

Just as mammals come in many shapes and sizes, so do their homes. Beavers often build dome-shaped lodges of sticks, stones, and mud. Chimpanzees make leafy nests in the trees. Many kinds of mammals dig underground burrows. Some take over dens made by others. Warthogs and pangolins move into abandoned aardvark burrows.

Underground homes can range from a single hole in the ground to a complex network of tunnels. Prairie dogs build burrows that contain rooms for nurseries, for sleeping, and for toilets. They even have rooms called listening posts. There they can listen for enemies above ground. A ring of earth at the mouth of the burrow helps keep water out.

Mammals that are active at night need shelters in which to spend the daylight hours. The flying squirrel, for example, builds a nest in a hollow tree.

Jackrabbits and other hares live in open areas. They often use bushes for shelter. As the hot summer sun moves across the sky, the animals stay in the shade by moving around the bush. Mammals like these, with no fixed homes to hide in, often have young that are independent soon after birth.

Some mammals such as caribou live in large herds that are continually on the move. Though a herd may not have one spot that is its home, it travels over a certain area as it looks for food.

The area in which an animal lives and feeds is called its home range. In regions where food is hard to find, an animal ranges over a large area.

On the next few pages, paintings show mammals of four climates and terrains. Though the animals shown in each illustration would not gather in the same place at the same time, they do inhabit the same general areas.

Polar bear wanders across the snow. These bears rarely seek shelter, except when they have young.

◁ Black-tailed jackrabbit hides from enemies and from the Texas sun in a shady spot under a bush.

Striped face and thick ▷ claws sandy from digging, a young badger sits at the entrance to its burrow. It makes its home in a riverbank in Michigan.

18

Northern flying squirrel peers from a hollow tree. Inside, a nest of shredded bark and small plants provides protection and warmth. Mammal homes range from simple shelters to complex ones.

ON THE DRY, OPEN GRASSLANDS of eastern Africa, large and small mammals—hunters and prey—gather at a water hole to drink and to eat. Most hoofed animals nibble bushes and grasses. Wild cats and wild dogs feed on meat. Others find insects or the remains of dead animals.

1. GIRAFFES
2. LIONS
3. ELEPHANTS
4. CHEETAH
5. IMPALA
6. LEOPARD
7. WILDEBEESTS
8. BURCHELL'S ZEBRAS
9. WATERBUCK
10. CAPE HUNTING DOGS
11. SPOTTED HYENAS
12. BABOONS
13. THOMSON'S GAZELLES
14. WARTHOGS
15. ROCK HYRAXES
16. DWARF MONGOOSES
17. BAT-EARED FOX

ALASKA'S COOL WATERS and stark, rocky coasts support many of the world's marine mammals. Whales and porpoises spend their entire lives in water. Seals, sea lions, and walruses go ashore to bear young.

1. HARBOR SEAL
2. SEA OTTER
3. NORTHERN SEA LION
4. PACIFIC WALRUS
5. NORTHERN FUR SEAL
6. HUMPBACK WHALE
7. RIBBON SEAL
8. HARBOR PORPOISE
9. ORCA
10. BAIRD'S BEAKED WHALE
11. DALL'S PORPOISE
12. BELUGA
13. GRAY WHALE
14. BLUE WHALE
15. BOWHEAD WHALE

IN THE DENSE FORESTS of South America, mammals live at different levels. Sloths and monkeys rarely come down from the treetops. A porcupine and an opossum clamber on lower branches. Such animals as agoutis and capybaras search for food among the tangled roots.

1. TWO-TOED SLOTH
2. SPIDER MONKEYS
3. UAKARI
4. KINKAJOU
5. EMPEROR TAMARIN
6. OPOSSUM
7. PREHENSILE-TAILED PORCUPINE
8. TAMANDUA
9. BROCKET DEER
10. JAGUAR
11. BRAZILIAN TAPIR
12. WHITE-LIPPED PECCARIES
13. CAPYBARAS
14. GIANT ARMADILLO
15. AGOUTIS

ALPS OF EUROPE *provide varied habitats for hoofed and fur-bearing mammals. Red deer and a marten take shelter in the woods. A marmot, an ermine, and a hare live among boulders and bushes. Ibexes and chamois find footholds on rocky slopes.*

1. CHAMOIS
2. ALPINE IBEXES
3. MOUFLON
4. ALPINE MARMOT
5. BLUE HARE
6. ERMINE
7. RIVER OTTER
8. BEECH MARTEN
9. RED DEER
10. WILD BOAR
11. LYNX
12. WOOD MOUSE

Finding Food

△ *Koala chews on eucalyptus leaves. These Australian animals feed in trees. Many kinds of mammals browse, or nibble on leaves and twigs.*

FOOD IS THE FUEL that keeps bodies working. But many different kinds of food—from berries and leaves to fish and meat—can provide the energy mammals need to survive.

Some mammals eat mainly plants and fruit. Others eat mostly meat. Still others, including humans, eat both plants and meat. Bears, for example, change their diet with the season. They may eat new shoots in the spring, fish in the summer, and nuts in the fall. At any time of the year, they may eat animals that they kill or find already dead.

To find food, most mammals probably rely on smell more than on their other senses. Squirrels sniff to find the nuts they have buried. Red foxes use their noses to detect the scent of nearby prey.

Other mammals use keen eyesight to spot prey. Cheetahs watch the grasslands of Africa for antelopes. When they see prey, they start to stalk, creeping closer and closer. Then they run it down. The serval, a kind of small cat, uses its ears to detect the rustling of small animals in the grass or underbrush. This hunter pounces on its prey.

Some bats use a very different system to find food. They hunt—and find their way—by echolocation (say ek-oh-low-KAY-shun). These bats send out beeps or pulses of high-pitched sounds. The bats listen for the sounds that bounce back when the beeps hit an object. From the echoes, they know where the object is and if it is moving.

Certain kinds of whales have still another way of gathering food. Inside the mouths of these whales

With a long tongue, a Cape pangolin licks ants from the surface of an acacia log. These scaly mammals also feed on termites. A pangolin tears open a termite mound and reaches in with its worm-shaped tongue. The insects stick to the tongue and the pangolin swallows them.

are baleen (say buh-LEAN) plates that serve as huge strainers. A whale opens its mouth to feed as it swims through water rich with shrimplike animals. Then it closes its mouth and pushes the water out with its tongue. Caught in the baleen are thousands of the tiny animals that make up the whale's meal.

Anteaters and pangolins reach their food by using large claws to tear openings in ant or termite nests. The long-nosed animals put their snouts against the holes and push their long tongues in. Many insects stick to their tongues and are swallowed. Anteaters and pangolins have no teeth at all.

Most other mammals have teeth that are specially suited for chewing the food they eat. Meat eaters usually have sharp teeth with which to seize prey and rip flesh. Animals that eat plants often have shovel-shaped teeth that cut off mouthfuls of food. Both kinds of animals have strong back teeth for grinding plants or for shearing meat from bone.

Many hoofed animals, like cows and deer, swallow their food without chewing it thoroughly. After eating, an animal brings up a wad of partly digested food, called a cud. It chews the cud thoroughly, swallows it, and digests it.

△ *Cheetahs approach a herd of wildebeests on the plains of Africa. If the cheetahs can get close to a small or weak animal—perhaps a young wildebeest—they will begin to chase their prey down.*

▽ *As a fishing bat skims above water, its hind claws break the surface. Fishing bats use echolocation to find prey. They use their large claws to hook it.*

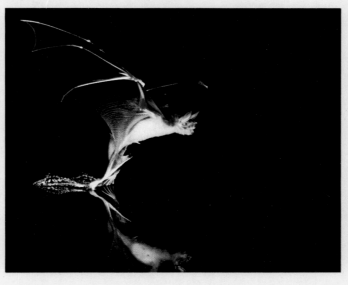

Getting From Place to Place

Leaping in unison, common dolphins speed through water off the coast of Mexico.

WALKING, HOPPING, crawling, swimming, or swinging, mammals travel from one place to another. Most mammals live on land and use all four limbs to move across the ground. Some, such as bears and raccoons, walk on the flat soles of their feet. Dogs and cats walk on their toes. Many fast-moving animals—horses, deer, and antelopes—run on hard toenails called hooves.

▽ *Over short distances, mammals travel at widely varying speeds. The sloth moves slowly through the trees at less than 1 mile (2 km) an hour. The armadillo ambles along more quickly. The cheetah, fastest land mammal, can reach speeds of 60-70 miles (97-113 km) an hour. The pronghorn comes close. A human sprinter can run a 220-yard, or 200-meter, race in less than 20 seconds—a rate of about 23 miles (37 km) an hour.*

1. Three-toed sloth: *less than 1 mile (2 km) an hour*

4. African elephant: *24 miles (39 km) an hour*

2. Nine-banded armadillo: *8 miles (13 km) an hour*

3. Gray squirrel: *12 miles (19 km) an hour*

5. Indian rhinoceros: *30 miles (48 km) an hour*

6. Red kangaroo: *30 miles (48 km) an hour*

What Is a Mammal?

Thick vine makes a natural jungle gym for a female ▷ orangutan and her young. Strong hands and long arms equip these primates for life in the trees.

Other swift mammals have developed strong back legs. Big kangaroos and little jerboas use their powerful hind limbs to bound across open areas.

Moles and pocket gophers live underground. These animals have powerful shoulders, short arms, and long hard claws, which they use to burrow easily through the soil.

When faced with water, most land mammals can swim if they must. Some do so especially well. Beavers and otters have webbed feet that help them swim. Mammals that spend all their lives in the water, like whales and manatees, have large flattened tails. They propel themselves along by moving their tails up and down.

Animals that live in trees often move easily among the branches. The sharp claws of opossums and squirrels help them climb. Monkeys cling to trees with strong hands. Binturongs and silky anteaters use their tails to help hold on.

Flying squirrels and flying lemurs glide through the air by spreading flaps of skin that connect their limbs. Of all the mammals, only bats can actually fly.

7. Giraffe: *35 miles (56 km) an hour*   8. Mexican free-tailed bat: *40 miles (64 km) an hour*   12. Cheetah: *60-70 miles (97-113 km) an hour*

9. Thoroughbred horse: *42 miles (68 km) an hour*   10. Black-tailed jackrabbit: *45 miles (72 km) an hour*   11. Pronghorn: *57 miles (92 km) an hour*

31

Defense and Offense

RUNNING AWAY from danger is often the best defense for most mammals. However, not all animals are fast enough or alert enough to avoid predators (say PRED-ut-erz)—the animals that hunt them. Sometimes mammals protect themselves by keeping perfectly still. As a predator passes by, it may not see or hear its prey.

Some mammals have coats that blend in with their surroundings. This kind of coloring is called camouflage (say KAM-uh-flazh). It may save an

animal from attack. Camouflage also may allow a predator to hide from its prey. The predator can stalk undetected until it is close enough to attack.

Some mammals find protection in groups. As a defense against wolves, musk-oxen stand shoulder to shoulder and lower their heads. The attackers face a wall of thick, sharp horns. Prairie dogs guard their groups of burrows. If a prairie dog spots danger, it barks a warning to the others.

If they are caught or cornered, or if their young are in danger, most mammals will put up a fight. Zebras and horses kick. Anteaters slash with their claws. Even tiny mice bite.

Some mammals have special body features or abilities that help them defend themselves. A porcupine's quills can prevent attackers from harming it. The strong scent of a skunk keeps enemies away. An opossum plays dead to discourage predators.

Teeth and claws, armor and quills all are used

△ Female hooded seal bares her teeth. Unlike many other seals, hooded seals fight to defend their pups.

for defense. Some of these features may be used in attacking other animals. Many mammals must be able to catch prey to get food.

Sometimes an animal's defensive features come in handy in several situations. In a tight spot, a walrus fights with its tusks. It also uses them in sparring with rivals and to nudge others away from its area. Sometimes a walrus's tusks even serve as hooks. The animal plants them on the ice and hauls itself out of the water.

◁ Wrestling match—more a game than a fight—erupts between two young male red kangaroos. The brief battle ended without injury to either.

▽ Spotted coat of a black-tailed fawn helps it hide among thick underbrush.

▽ Two sassabies stand alert on a termite mound. From this raised area on a grassland in Africa, these antelopes can spot an approaching enemy.

Caring for Young

What Is a Mammal?

BECAUSE IT MUST FEED on her milk, a newborn mammal depends on its mother. Some young, however, need more attention than others.

Newborn marsupials are especially helpless. The kangaroo, the koala, the opossum, and the other members of this order all bear tiny, underdeveloped young. An offspring crawls into its mother's pouch soon after birth. There it remains attached to its mother's nipple. Only after several weeks is it large enough to move around—even in the pouch.

Mice, wolves, rabbits, and many other animals are born blind and helpless. They must nurse, and they need a long period of care. The mothers of these animals can leave their offspring alone for short periods while hunting for food. In contrast, hares, wildebeests, and zebras are relatively independent at birth. Porcupines can walk when they are only a few minutes old. Cavies nibble plants soon after birth.

Some mammals cooperate in caring for young. A female lion may nurse another's cubs. Cape hunting dogs share food with each other's pups.

Some kinds of mammals, such as chimpanzees, have complex social systems that shape the ways they behave. By copying adult behavior, the young of these species learn the skills they need to survive and to live in groups. The young learn how to find food, how to groom themselves and each other, how to build nests, and how to behave toward other members of their group.

▽ *After a meal, a female cheetah grooms her purring five-month-old cub by licking its face.*

△ Lending a helping trunk, an African elephant tugs a calf out of the water. Young elephants can walk soon after birth. But during the first months of their lives, they often need the help of adults.

Inside a lodge of sticks, stones, and mud, a female beaver nurses △ one of her furry twins.

Two female rhesus monkeys groom their young. This social activity ▷ helps keep hair and skin clean.

35

Sounds and Signals

△ Twin pronghorn young bound after their mother. Their white rump patches serve as signals that danger lurks nearby. Other pronghorns that see the signal may also take flight.

△ Young hoary marmot touches noses with an adult. These rodents often greet each other this way.

Head thrown back and body held erect, a black-tailed ▷ prairie dog calls out shrilly, possibly to defend its home. Sometimes an excited prairie dog may jump into the air or even fall over backward. If this North American rodent spots danger, it barks an alarm.

AS A PRONGHORN dashes across the prairie, the white hairs of its rump stand on end. This flash of white can be seen for a long distance. It serves as a signal to other pronghorns that danger is near. This kind of signal is only one of the many ways that mammals communicate with each other.

Communication does not mean that animals talk to each other as people do. Mammals communicate in several silent ways. For example, they use face and body positions to communicate. A chimpanzee lets others know how it feels by the expression on its face. The leader of a wolf pack usually carries his tail straight out. All the other members of the pack walk with their tails drooping. After a fight, one wolf indicates that it recognizes another as leader by lying down and showing its throat.

Many mammals communicate by the positions of their ears. Ears laid back may mean that an animal is annoyed or angry. Perked ears may show that an animal is alert.

Scent helps some mammals find each other, and it keeps others apart. Such animals as cats and dogs mark territories by spraying urine on trees and rocks. These marked places are called scent posts.

Many other animals also leave scent marks. Rabbits and hares produce a strong-smelling substance in glands under their jaws. They rub their jaws on the ground throughout their home areas. During the mating season, male pronghorns mark trees and bushes with scent from glands near their eyes. Both scent and claw marks on trees announce a bear's presence to other bears.

Some mammals use sounds to communicate. They make noises to frighten enemies, to attract mates, and to challenge rivals. They use sounds to call their young, to signal that danger has passed, or to gather others of their kind.

During the mating season, many mammals are noisier than usual. A bull elk has a strong, deep voice. His bugling sounds may attract females and keep rivals away.

To sound an alarm, a marmot whistles. A beaver warns of danger by striking the surface of the water with its tail. This loud whacking sound carries a long distance. A raccoon makes low throaty noises as it moves about. If danger threatens, the female's "churr churr" becomes louder and louder, until she screams—sending her young for cover.

Crossing snowy terrain in Michigan, a group of five gray wolves hunts for prey. The dominant wolf, or leader (second from last), walks with his tail held straight out. The tails of the other wolves droop. This may show that they recognize the dominant wolf as the most important member of the pack.

Season by Season

IN A FEW ENVIRONMENTS around the world—such as tropical rain forests—the climate changes little from one season to the next. Food is usually available all year long. The lives of mammals that inhabit these areas also remain much the same whatever the season. Where the seasons do change, however, mammals must adapt their behavior. They must be able to get food and shelter in different kinds of weather.

Many mammals store food in the summer and fall. Rodents often put away nuts and seeds in their burrows. During the winter, when fresh food is scarce, they eat what they have saved. The pika builds haystacks. It gathers herbs and grasses and spreads them out to dry. Then it piles the dry food under rock ledges for use during the winter.

Some mammals, such as marmots and dormice, get along during cold weather by means of hibernation (say hye-bur-NAY-shun). A hibernating animal eats a great deal during the summer and fall and grows fat. When winter comes, it crawls into a burrow or den and drops into a kind of sleep. Its body temperature goes down, and its breathing and heart rate slow. It uses only a small part of the energy it

△ *Curled nose to tail, a golden-mantled ground squirrel sleeps during the winter. While it hibernates, its temperature drops and its breathing and heart rate slow.*

◁ *Crew-cut summer coat appears as a mountain goat sheds its winter hair. New hair will grow out all summer, forming a long, thick coat by fall.*

would need if it were active. It lives on the fat in its body. Animals that behave similarly in hot or dry weather are said to aestivate (say ES-tuh-vate).

Mammals that live in cool climates usually have two different coats of hair. Their thick winter coat serves as insulation, keeping body heat in and cold out. This heavy coat is shed in warm weather. It is replaced by a thinner summer coat. This process is called molting.

A few mammals, such as the snowshoe hare and some weasels, change color as the seasons change. In summer, their hair is brown. Gradually,

△ *As spring arrives, male bighorn sheep feed on plant shoots in Glacier National Park in Montana. All winter, they nibbled plants on lower slopes. When summer comes, they will move higher.*

◁ *In summer, a short-tailed weasel's brownish coat (left) blends well with rocks. Some weasels change color with the seasons. A white-coated weasel (far left) is hard to see against the snow.*

new white hair grows in. By winter, the summer hair has been shed, and the coats are white. The color changes help camouflage the animals. Their white coats blend in with snow. Their brown coats are hard to see against rocks and dry grasses.

Some mammals escape harsh winters by migrating, or traveling, from one place to another. In the summer, mountain sheep feed high in the mountains. They come down to lower areas during cold weather. Certain whales swim to warmer water in winter. They spend only the summer months in the polar regions.

Studying Mammals

Dr. Merlin Tuttle examines a frog-eating bat lured into a net by recorded frog calls.

WHY STUDY MAMMALS? What makes us want to learn more about this varied group? We share the earth with mammals. We use them as sources of food and clothing, as work animals, and as pets. We too are mammals. By studying other mammals we learn more about ourselves.

Scientists who study mammals are called mammalogists (say muh-MAL-uh-justs). Mammalogists may be interested in certain specialized subjects. The study of anatomy (say uh-NAT-uh-mee) tells how mammals' bodies are put together. Physiology (say fizzy-AHL-uh-gee) is the study of how mammals' bodies work. Taxonomy (say tak-SAHN-uh-

mee) classifies the many kinds of mammals and reveals how they are related. Some mammalogists learn about ethology (say ee-THAHL-uh-gee), or the study of why mammals behave as they do. Ecology (say ee-KAHL-uh-gee) teaches how mammals interact with their environments.

Some scientists study mammals in laboratories. They may experiment with live animals in carefully controlled environments. Or they may use collections of preserved mammals or mammal skeletons. By measuring mammals and comparing sizes and other differences, scientists can tell a great deal about how different species have developed. Other kinds of scientists work where mammals live in the wild. To study mammals, scientists use many kinds of equipment, from tape recorders and cameras to detailed journals, collection boxes, and special measuring instruments.

But you don't have to be a mammalogist to study mammals in some way. Most of us would find it too hard to study such animals as giraffes and aardvarks in the wild. Squirrels, chipmunks, and raccoons, however, often live near people. We can see what they look like and how they behave. Pet dogs and cats also are easy to study. We can learn about mammals by looking at the animals around us and by watching those that share our lives.

Male caribou carries a radio transmitter through a park ▷ *in Alaska. Signals from such collars permit rangers to track the animal's movements.*
▽ *Along a creek in South America, mammalogist Nicole Duplaix measures waste left by giant otters. By studying these traces, scientists can learn much about mammals in the wild.*

▽ *Artificial burrow in a laboratory allows scientists to study naked mole rats.*

41

Survival in Question

HUMAN BEINGS share the earth with all other mammals. People change their environment in many ways to suit themselves. They often do not think about how these changes affect other animals. They build houses for shelter, plow the land to plant crops, and use natural resources for energy. In some places, human activity destroys the homes of other kinds of mammals.

The mountain gorilla, for example, lives in a small area in Africa. Even though the area is part of a park, local people increasingly farm in the area. As the gorillas' habitat gets smaller and as animals are killed by poachers, scientists worry that the species may die out.

In many parts of the world, such as Australia, people have introduced, or brought in, animals that had never lived there before. Many native animals could not compete with the newcomers for food—or they themselves became prey. Some of them died out completely.

Over long periods of time, other species have not been able to adapt to natural changes and have

also become extinct. As these animals have disappeared, other species have replaced them.

Species that are in danger of extinction now can be saved, or at least protected. One way to do this is to set aside special areas for wild animals. On wildlife preserves and in zoos, many endangered species have another chance for survival.

One animal that has been rescued in this way is the Arabian oryx. Once there were only a few of these graceful animals left in their desert homeland. In the early 1960s, some were captured and taken to

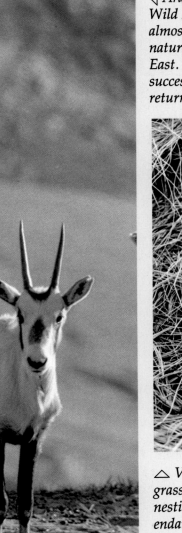

◁ *Arabian oryxes live at San Diego Wild Animal Park. The animals had almost disappeared from their natural environment in the Middle East. These oryxes have bred so successfully that some may be returned to the wild.*

△ *Volcano rabbit crouches in the grass, which it uses as food and as nesting material. These rabbits, endangered by a shrinking habitat, live only on the slopes of a few volcanoes in Mexico.*

△ *Golden lion tamarins huddle on a tree branch. These tiny, endangered animals survive in captivity. The destruction of their natural habitat in Brazil means they probably will never again live in the wild.*

preserves in Africa and in the United States. There the oryxes have adapted to new environments, have bred, and have raised young. Now there are enough Arabian oryxes that some may be returned to their natural environment.

For many species of mammals, conservation in zoos and breeding in captivity offer the only hope of survival. Golden lion tamarins, sometimes called marmosets, survive mainly in zoos. Their natural habitat—the rain forests of a small area of Brazil—has been nearly destroyed. These animals will probably never be taken back to the wild.

Scientists continue to study ways to ensure the survival of endangered mammals. Some governments and individuals help by taking an active interest in conservation. People can try to make sure that their descendants will see and enjoy the other mammals that inhabit earth's deserts and mountains, its seas and forests, its swamps and grasslands.

A

Aardvark

(say ARD-vark)

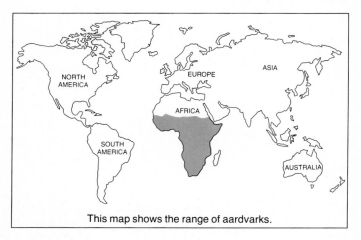

This map shows the range of aardvarks.

Long ears and piglike snout of an aardvark cast a shadow on a termite mound in Africa. Keen hearing and a sharp sense of smell help the animal find food and avoid danger. Above, strong front claws make dirt fly as an aardvark burrows into the side of a mound in search of insects to eat.

THE SPELLING OF ITS NAME gives the aardvark its place at the beginning of most animal lists. Aardvark means "earth pig" in the Afrikaans language of South Africa. Its snout does look like a pig's. But this long-nosed, long-eared animal is not related to the pig or to any other mammal!

The aardvark's nighttime habits make the animal difficult to find and to study. Only after sunset does it leave its burrow in the grasslands or the forests to search for food. Outside the den, the stocky animal, which measures about 6 feet (183 cm) from head to tail, pauses. It listens and sniffs for danger. If all is safe, it trots away. The aardvark moves on its toes and claws, often following a zigzag path. Its tail drags behind, making a groove in the ground.

When the aardvark finds a termite mound, like the one in the picture at right, it digs a hole near the base. The sunbaked earth of a mound can dry as hard as concrete. But the aardvark is a strong burrower. With the thick, sturdy claws of its front feet, it can burrow through even hard-packed soil. The aardvark then pushes its blunt snout close to the opening in the mound. It catches the termites with its long, worm-shaped tongue.

Tough skin protects the aardvark from insect bites. The animal can even close its nostrils, so that termites, ants, and dust do not get into its snout.

After the aardvark has eaten from one mound, it may move on to another or dig into an underground ant nest. Aardvarks may travel several miles a night searching for food.

Although it is a timid animal, the aardvark can fight off attackers such as big cats and wild dogs. It sits on its rump and lashes out with its front claws. Sometimes the aardvark lies on its back and slashes at an enemy with all four feet. But rather than fight, the aardvark will try to escape from danger. It runs for its den or quickly digs another.

By morning, the aardvark returns to its cool tunnel-like burrow. All day, the aardvark sleeps there curled up in a circular room. There is just enough room for the aardvark to turn around—and leave its den headfirst.

A female aardvark usually has one offspring a year. The hairless newborn has tender, pinkish skin. It stays in the den for about two weeks. Then it begins to search for food with its mother. After six months, the young aardvark digs its own burrow. But it stays near its mother for several months more.

AARDVARK

LENGTH OF HEAD AND BODY: 43-53 in (109-135 cm); tail, 21-26 in (53-66 cm)

WEIGHT: 110-180 lb (50-82 kg)

HABITAT AND RANGE: forests and grasslands of central, southern, and eastern Africa

FOOD: usually termites, ants, and some fruit

LIFE SPAN: about 10 years in captivity

REPRODUCTION: 1 or 2 young after a pregnancy of about 7 months

ORDER: tubulidentates

Aardwolf

AARDWOLF

LENGTH OF HEAD AND BODY: 22-31 in (56-79 cm); tail, 8-12 in (20-30 cm)

WEIGHT: 20-31 lb (9-14 kg)

HABITAT AND RANGE: open woodlands and plains in eastern and southern Africa

FOOD: mostly termites

LIFE SPAN: 14 years in captivity

REPRODUCTION: usually 2 to 4 young after a pregnancy of about 3 months

ORDER: carnivores

AT NIGHT, IN THE WOODLANDS and on the dry plains of southern and eastern Africa, the aardwolf searches along the ground for termites. When it finds food, the animal quickly laps the insects up with its long, sticky tongue. Then the aardwolf may groom itself. It uses its tongue to clean its narrow muzzle and the inside of its mouth. It lies down and licks its striped, yellowish fur.

Though its name means "earth wolf" in the Afrikaans language of South Africa, the aardwolf is not related to the wolf. The seldom-seen animal belongs to the same family as the hyena. Aardwolves are sometimes mistaken for striped hyenas, though they are smaller. They grow only about 3 feet (91 cm) long from head to tail. Aardwolves are carnivores, that is, they are meat eaters. But their jaws and teeth, unlike those of their relatives, are too weak to crush anything but insects. Find out more about hyenas on page 278.

During the day, aardwolves sleep in holes in the ground. Usually, they stay in old burrows left by aardvarks or other animals. Occasionally aardwolves dig dens of their own. They may leave them in the morning to lie in the sun. If the weather is bad, they may stay in their dens for days at a time.

A female aardwolf usually gives birth in a den. The newborn cubs—normally three in a litter—are blind and helpless. They are fully grown in about nine months.

△ *From ears to bushy tail, an aardwolf stands alert on the dry plains of Africa. If threatened, the animal will make itself look larger by raising its dark mane.*

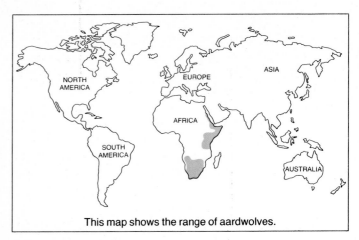

This map shows the range of aardwolves.

◁ *Aardwolf licks up termites with its sticky tongue. It quickly feeds on insects from one area and then moves on. The animal can eat about 40,000 termites in three hours!*

When threatened by other animals, an aardwolf raises its black-and-yellow mane. This ridge of hair extends down the neck and back. When the mane stands on end, the animal looks much larger

Agouti

(*say* uh-GOOT-ee)

AGOUTI

LENGTH OF HEAD AND BODY: 16-24 in (41-61 cm); tail, about 1 in (3 cm)

WEIGHT: 2-9 lb (1-4 kg)

HABITAT AND RANGE: tropical forests in Mexico and in Central and South America

FOOD: fruit, leaves, roots, and stalks

LIFE SPAN: 13 to 20 years in captivity

REPRODUCTION: 1 or 2 young after a pregnancy of about 3 months

ORDER: rodents

Thirsty agouti heads for the water's edge during the dry ▷ *season in Bolivia. Quick and alert, it freezes when it hears a noise. If discovered by an enemy, it will dash for cover.*

SITTING UP AND LISTENING to the sounds of the forest, an agouti holds an avocado between its forepaws. It peels the soft fruit with its teeth and eats it. Suddenly, the agouti perks its short ears. Sensing danger, it freezes. A rustle of leaves warns of an approaching enemy—perhaps an ocelot.

With a call of alarm, the agouti dashes through the forest to escape its enemy. It swiftly zigzags among the trees. Using its strong legs, the rodent tries to outrun the wild cat. Or the agouti tries to trick its pursuer. It darts into a hollow log and slips out the other end.

There are about ten kinds of agoutis. The animals have short tails and small, rounded ears. Agoutis make their homes in the tropical forests of Mexico, Central America, and South America. Usually they live in sheltered spots under tree roots, between rocks, or in hollow logs.

During the day or in the evening, agoutis look for parts of plants to eat. Their coarse hair—pale orange, brown, or almost black—blends with the colors of the forest. Though an agouti's coat appears to be one shade, each hair has bands of color.

than usual. The aardwolf also barks and roars ferociously. Though the aardwolf rarely fights, it will use its sharp, small teeth if it finds itself cornered. Usually it hides in its den.

Female agoutis usually bear one or two young. The newborn are covered with hair and are more developed than the offspring of some other rodents. Able to see at birth, they often nibble on green plants an hour later. Like full-grown agoutis, they freeze when in danger.

A close relative of the agouti is the paca. You can read about the paca on page 426.

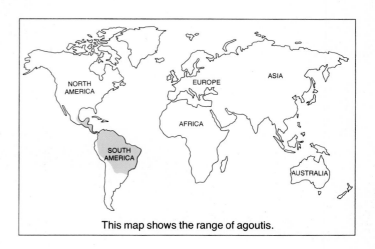

This map shows the range of agoutis.

Alpaca
The alpaca is a close relative of the llama. Read about alpacas and llamas on page 342.

Anteater

(say ANT-eat-er)

Giant anteater: 49 in (124 cm) long; tail, 35 in (89 cm)

Holding onto its mother's back, a five-month-old giant anteater hitches a ride in the grasslands of Brazil. Broad, dark stripes on the animals' sides line up, helping to hide the young against the adult's fur.

DOES AN ANTEATER EAT ANTS? Yes, this remarkable animal with the long snout really lives up to its name. To reach its food, the anteater scratches a hole in an anthill with a sharp, curved claw. It darts its long tongue through the hole and inside the nest. With a flick, it jerks back its tongue, which is covered with ants. The insects are swallowed whole.

An anteater does not linger at an anthill. It may stay for less than a minute. The longer it feeds, the more chance the insects have to sting.

The anteater eats several kinds of insects. But ants and termites are the animal's main foods. In fact, an anteater can eat thousands of ants and termites in a single day!

48

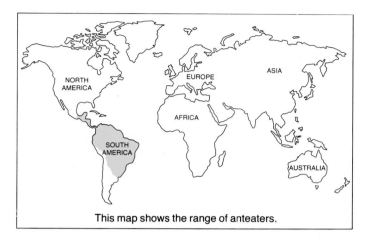
This map shows the range of anteaters.

ANTEATER

LENGTH OF HEAD AND BODY: 6-49 in (15-124 cm); tail, 7-35 in (18-89 cm)

WEIGHT: 6 oz-86 lb (170 g-39 kg)

HABITAT AND RANGE: tropical forests and grasslands from southern Mexico through northern Argentina

FOOD: mostly ants and termites

LIFE SPAN: 3 to 14 years in captivity, depending on species

REPRODUCTION: usually 1 young after a pregnancy of about 5 or 6 months

ORDER: edentates

An anteater's mouth is nothing more than a pencil-size hole at the end of its snout. The animal has no teeth, so it cannot chew the insects it eats. The ants and termites are crushed and digested in the anteater's stomach.

The anteater does not see very well. It depends on its nose to lead it to an insect nest. Its sense of smell is much better than a person's.

Anteaters live in tropical forests and grasslands of Central and South America. The squirrel-size silky anteater stays mostly in trees. It gets its name from its soft, silky fur. The raccoon-size tamandua (say tuh-MAN-duh-wuh) divides its time between the trees and the ground. The tamandua is sometimes called a collared anteater because of the ring of light-colored fur around its neck.

At night, both the tamandua and the silky anteater look for food. They move slowly but easily through the branches, grasping them with their long tails and hooklike claws. During the day, they curl up into tight balls and sleep in tree forks.

The giant anteater measures as long as 7 feet (213 cm) including its tail. It rarely climbs trees. It spends most of its waking hours sniffing for food on the ground. It seems to shuffle along on its front knuckles. Actually, it is walking on the sides of its paws. Unlike a cat, the giant anteater cannot pull in its strong claws. To protect them and keep them sharp, it curves them under its body as it walks.

The giant anteater usually stays in its home territory. It takes to water easily, and sometimes it even swims across wide rivers in search of food. When it tires, the animal lies down and covers itself with its bushy tail. The tail serves as a blanket on cold nights and helps to hide the animal from its enemies.

Anteaters usually live alone, but females are sometimes seen with young. An anteater gives birth about once a year to a single offspring. A young anteater spends much of its time riding on its mother's back. A young giant anteater often becomes hidden in its mother's thick fur. When it is older, it occasionally will gallop alongside its mother.

A female silky anteater may leave her young hidden in a tree nest of dry leaves while she looks for food. When the young is asleep, it blends in with the branches, and enemies cannot easily see it.

Though anteaters never attack another animal first, they will defend themselves fiercely. When in danger, the giant anteater strikes out with its thick, strong claws. Sometimes it even rears up on its hind legs. A giant anteater's claws measure as long as 4 inches (10 cm). The animal is a match even for a mountain lion or a jaguar.

Long, tapered snout helps a giant anteater sniff for food. The animal has such poor eyesight that it probably could not see a person standing just a few feet away.

49

Silky anteater: 6 in (15 cm) long; tail, 7 in (18 cm)

◁ *Carrying her offspring on her back, a female silky anteater walks along a branch in an Amazon forest. This kind of anteater spends almost all of its time in trees.*

Rising on its hind legs, an alarmed tamandua prepares ▷ *to defend itself. The animal's sharp, curved claws can cause serious wounds if an attacker gets too close. Its tail acts as a prop. Another tamandua (far right) uses its tail to hold on as it climbs down a tree in South America.*

▽ *Giant anteater feeds on insects at a termite mound. The animal flicks its worm-shaped tongue in and out as many as 160 times a minute. It catches thousands of ants and termites every day.*

Tamandua: 21 in (53 cm) long; tail, 21 in (53 cm)

The tamandua uses its claws as weapons, too. When it is startled, it rises on its hind legs and spreads out its paws. If there is no real danger, it will drop back on all fours and move away. If attacked, however, the tamandua may strike with its razor-sharp claws. Or it may grab its enemy in a strong grip and hold it away from its body until the attacker is stunned or dead.

Like the other anteaters, a threatened silky may rise on its hind legs to defend itself. It braces itself with its tail. Then it raises its front paws above its head and strikes down hard, slashing at its enemy. But a silky anteater's best defense is camouflage. Its gray-brown fur is hard to see against the branches in which it lives.

Anteaters are fairly common in parts of South America. Laws in Brazil protect the giant anteater. But elsewhere it is still hunted by people. It is an easy target and can be seen from far away, lumbering across the grasslands.

51

Antelope

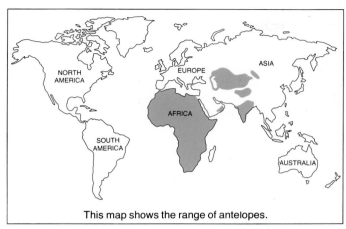

This map shows the range of antelopes.

ANTELOPE

HEIGHT: **10-71 in (25-180 cm) at the shoulder**

WEIGHT: **4-2,000 lb (2-907 kg)**

HABITAT AND RANGE: **many kinds of habitats in Africa, central and southern Asia, and the southwestern Soviet Union**

FOOD: **grasses, herbs, leaves, twigs, bark, buds, fruit, and insects**

LIFE SPAN: **3 to 25 years in captivity, depending on species**

REPRODUCTION: **1 to 3 young after a pregnancy of 4 to 9$\frac{1}{2}$ months, depending on species**

ORDER: **artiodactyls**

MANY PEOPLE think of antelopes as handsome, deerlike animals that bound gracefully across the plains of Africa. Some antelopes do fit this description. But this group of hoofed animals is amazingly varied. From the jackrabbit-size royal antelope to the oxlike eland (say EE-lund), antelopes come in a wide range of shapes and sizes. A look at their horns gives a good idea of how different these animals can be. Horns may be straight, curved, twisted, spiraled, and ringed. Each of the nearly one hundred species, or kinds, of antelopes has a uniquely shaped set of horns.

Some kinds of antelopes are found in Asia—the saiga (say SIGH-guh) and the black buck, for example. But most live in Africa. They are found in nearly

Peaceful scene at water's edge: Several springboks share a water hole with wading birds. The rest of the herd grazes behind them. Dutch settlers in South Africa named these antelopes for their habit of jumping straight up into the air when startled.

Springbok: 30 in (76 cm) tall at the shoulder

53

Antelope

Horns crossed, two gemsboks in Namibia clash in ▷ battle while another watches. Gemsboks may fight for mates or to test their strength. Each twists and turns its head, as if trying to throw its rival to the ground. The stronger animal wins. Gemsboks, like most antelopes, rarely hurt each other when fighting. But they do use their swordlike horns as weapons against lions and other large predators.

▽ Female defassa waterbuck and their month-old young jump into a lake in Kenya. True to their name, these large antelopes seldom roam far from water.

Gemsbok: 48 in (122 cm) tall at the shoulder

Defassa waterbuck: 48 in (122 cm) tall at the shoulder

Saiga: 30 in (76 cm) tall at the shoulder

Bulky noses and bulging eyes identify saigas (above) of the southwestern Soviet Union and central Asia. Their large nostrils may warm and moisten cold, dry air. Huge ears erect, a young female greater kudu (right) stands near a water hole in Namibia. Greater kudus can go several days without drinking.

54

Greater kudu: 59 in (150 cm) tall at the shoulder

Scimitar-horned
oryx: 47 in (119 cm)
tall at the shoulder

Klipspringer: 20 in (51 cm) tall at the shoulder

△ *Graceful curving horns sprout from the heads of scimitar-horned oryxes. Both males and females grow horns about 3 feet (91 cm) long. People sometimes hunt these desert antelopes for meat and for their horns.*

Surefooted klipspringer perches on a rock. ▷ *This small antelope lives in rocky and mountainous regions of eastern, central, and southern Africa. Dutch settlers named the animal "cliff springer" because it leaps easily from rock to rock. The klipspringer's rubberlike hooves help it land safely on rock ledges no bigger than a teacup. Like a dancer on tiptoe, it stands and walks on the very tips of its hooves.*

every kind of habitat—deserts, swamps, and mountains. But most of the species are found in forests and on grasslands in Africa.

A small antelope like the duiker (say DYE-ker) usually lives in forests or in thick brush. Duikers feed on the leaves and twigs of bushes and trees. They may live alone, in pairs, or in small family groups. A male often protects the territory in which he and a female live, perhaps with offspring. If other male duikers try to invade this area, he chases them away. Since only a few duikers live in a territory, there is plenty of food to go around. Other antelopes may not have this kind of territory. And they may have to roam great distances in search of food.

Large antelopes such as elands often travel in herds of as many as 200 animals. In the rainy season, they gather on the plains. In the dry season, they scatter into brushy areas to find food and water.

Often several kinds of antelopes may roam the same area. Some graze, or eat only grass. Others browse, or eat only leaves and twigs. Still others do both. But the different kinds of antelopes do not compete for the same food. Smaller antelopes feed on the lower leaves of a plant. Larger antelopes eat the leaves higher up.

Most antelopes eat during the cool parts of the day—early morning and late afternoon. As they feed, the animals watch constantly for enemies. They eat quickly, swallowing their food nearly whole. Later, when resting, they bring up a mouthful of partly digested food—called a cud—and chew it thoroughly. Antelopes are related to other cud-chewing animals, such as camels, cows, deer, goats, and sheep. You can read about these animals under their own headings.

Lions, cheetahs, hyenas, wild dogs, and about

Red hartebeest: 49 in (124 cm) tall at the shoulder

△ *All four hooves off the ground, a red hartebeest gallops across an open plain in Botswana. Hartebeests live in herds of five to twenty animals. If an enemy approaches, the antelopes speed away in single file.*

▽ *Young suni sniffs curiously at its mother's mouth. Young antelopes often learn what they can eat by imitating their mothers. These small antelopes live in dense brush in eastern and southern Africa.*

Suni: 14 in (36 cm) tall at the shoulder

Zebra duiker: 16 in (41 cm) tall at the shoulder

a dozen other meat-eating animals are predators (say PRED-ut-erz), or hunters, of antelopes. People, too, hunt them for their meat, hides, and horns.

Antelopes have keen senses, and they are always alert to enemies. Sniffing the air, antelopes may pick up the scent of a predator. Their large ears hear the slightest sound. With huge eyes on the sides of their heads, they keep a sharp lookout for danger. When threatened, a small antelope, such as a dik-dik, may hide. A big antelope—an eland, for example—may try to stand its ground. But for many antelopes the best defense is speed.

Most antelopes are fast runners and can escape a predator in long leaps. Antelopes are built for swift movement. They have long legs, strong hindquarters, and angled hooves that help them jump.

Hiding, protective coloring, and traveling in herds are useful defenses. A small antelope such as the suni (say soo-nee) usually lives in areas that offer thick cover. If a jackal or other predator approaches, the suni hides among the bushes and tall grasses. It drops down and lies very still until the jackal passes. But if a jackal gets too close, the suni jumps up and runs away. When it gets far enough ahead of the predator, it drops down again and hides.

Springboks are too big to hide. They often rely on a different method to avoid enemies. When a predator comes close, these fleet-footed antelopes jump straight up into the air several times. As they leap, a patch of white hair flashes under their tails. This signal warns other springboks that danger is near. Then the animals run away.

Some antelopes, such as bongos and reedbucks, have coats that blend into their surroundings and act as camouflage.

Herds offer antelopes protection from predators because of the number of animals in one place. With so many eyes, ears, and noses alert, a lion or a cheetah has more difficulty sneaking up on a herd. Sometimes several members of a topi or a hartebeest herd act as guards. If they spot a predator, they snort in alarm and (Continued on page 60)

◁ *After a chilly night, a zebra duiker warms itself in the morning sun. Duikers usually feed after dark. Some kinds grow only as big as hares. Others grow as tall as deer.*

"Are you mine?" a female topi asks with a sniff. A female knows her offspring by its scent. She can find her young even in a large herd. A newborn can walk soon after birth.

Topi: 48 in (122 cm) tall at the shoulder

Black buck: 32 in (81 cm)
tall at the shoulder

Bongo: 48 in (122 cm) tall at the shoulder

△ Group of black buck chew grass on a preserve in India. There the animals live protected from hunters who want their meat and their horns.

◁ Broad stripe stretches across a young male bongo's face. The markings on its sides and back make the animal hard to see among forest shadows.

▽ Oxlike in size and appearance, an eland strides through the grass. Both male and female elands, the largest antelopes, grow horns. An old bull like this one often makes a clicking sound when walking. Scientists think the clicking lets other elands know of its presence.

Eland: 70 in (178 cm) tall at the shoulder

Bohor reedbuck: 30 in (76 cm) tall at the shoulder

△ *Male bohor reedbuck in Kenya watches for danger. It will dash away if an enemy approaches.*

▽ *Male nyala drinks at a water hole. When the white hair on its back bristles, the animal is signaling a threat.*

Kirk's dik-dik: 14 in (36 cm) tall at the shoulder

Nyala: 45 in (114 cm) tall at the shoulder

△ *Tiny Kirk's dik-dik, about 14 inches (36 cm) tall, becomes hard to see among grasses. A dark spot at the corner of the animal's eye produces a sticky substance. The dik-dik uses this to mark its territory.*

59

Beneath misty mountains, a herd of Uganda kobs grazes quietly on a game preserve in Africa. These animals feed mainly on grasses as they roam the open plains. When food is plentiful, hundreds of animals may gather to feed. Much of the year, kobs live in three separate groups. Females and young form a nursery herd. The strongest males guard their own territories. Males without territories stay together in a bachelor herd.

Uganda kob: 36 in (91 cm) tall at the shoulder

gallop away. Then the entire herd takes off. Often it is the slower-moving antelopes—the old, sick, or very young—that get caught by predators.

A few kinds of large antelopes, such as the gemsbok (say GEMZ-bahk) and the eland, use their horns as weapons against predators. Hard, hollow horns grow around two bony cores on an antelope's head. The horns keep growing throughout the animal's life. They do not fall off every year as a deer's antlers do. Not all female antelopes grow horns, but all males do. Males often fight with their horns.

Fighting among male antelopes is a show of strength, a contest to prove which antelope is stronger. The fights rarely end in death. In fact, antelopes usually do not draw blood when fighting. Dik-diks, one of the smallest kinds of antelopes, battle without even touching. Instead, male dik-diks charge at

each other as if they were going to attack. Then they stop short. They do this over and over until one dik-dik is forced back into its own territory.

Some antelopes mate at any time of the year. Others mate only during a certain season. At that time, fighting among male antelopes increases. A male kob, for instance, stakes out a territory. If another male comes into this area, the holder of the territory will fight it. Locking horns, the two push and shove. They continue fighting until one kob tires and leaves. The winner remains in the territory and mates with any female kobs that enter it.

Female antelopes usually bear one young at a time. Scientists group these young antelopes into hiders and followers. Smaller antelopes often hide out after they are born. The female gives birth to her offspring in a secluded spot safe from predators.

Usually the newborn lies hidden under a bush or in the grass for several weeks. The mother feeds as usual, but she returns to the hiding place several times a day to nurse her young. As the young grows bigger, it starts to wander with its mother. Within six months, it may become independent.

Offspring born to larger antelopes that roam in herds may be followers. The young can stand and walk shortly after birth. It stays close by its mother's side and goes with her everywhere. In a few days, it can run as fast as an adult antelope.

Read about some of the best-known antelopes under their own headings. You can find out about gazelles on page 212, gerenuks on page 220, impalas on page 292, and wildebeests on page 577.

Armadillo

Tough scales shield the body of a nine-banded armadillo, the only kind found in the United States. Thick skin and a few coarse hairs cover the unarmored parts of its head and legs.

Nine-banded armadillo: 18 in (46 cm) long; tail, 12 in (30 cm)

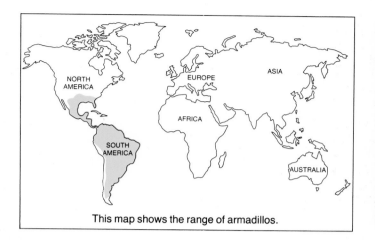

This map shows the range of armadillos.

△ *Young nine-banded armadillo and its mother sniff the ground in search of food. The young animal has soft leathery skin. As it grows, its skin will harden.*

▽ *With its snout in the soil, a nine-banded armadillo eats worms and insects. A keen sense of smell helps the armadillo find food, and a long tongue helps catch it. Armadillos often make grunting and sniffling noises as they search for a meal.*

ARMADILLO

LENGTH OF HEAD AND BODY: **5-37 in (13-94 cm); tail, 1-21 in (3-53 cm)**

WEIGHT: **3 oz-120 lb (85 g-54 kg)**

HABITAT AND RANGE: **grasslands and open forests from the southern United States through most of South America**

FOOD: **ants, termites, worms, snails, beetles, roots, fruit, snakes, and dead animals**

LIFE SPAN: **4 to 16 years in captivity, depending on species**

REPRODUCTION: **1 to 12 young; pregnancy varies by species and is not known for all**

ORDER: **edentates**

LIKE A SUIT OF ARMOR, plates of skin-covered bone protect most of the armadillo's body. *Armadillo* means "little armored one" in Spanish. The large, solid plates are connected by overlapping bands that partly circle the animal's middle and allow it to bend. The armor protects the armadillo from thorns and branches and sometimes from its enemies.

Armadillos are well equipped for digging. The stocky animals have short front feet with powerful, curved claws. They use these to dig underground

Nine-banded armadillo cuts through the water of a river. Armadillos swallow air as they swim. This helps them float. Sometimes an armadillo may cross a river by holding its breath and walking on the bottom.

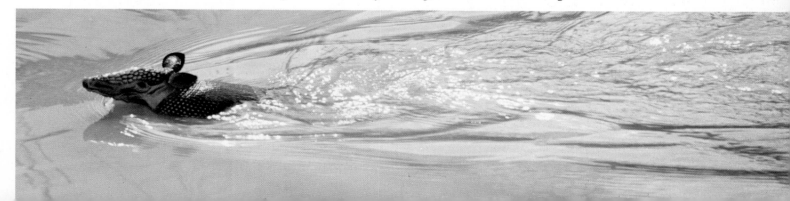

burrows where they sleep and raise young. Because an armadillo can hold its breath as long as six minutes, it does not breathe in dirt when it is digging. The nine-banded armadillo can even cross rivers by holding its breath and walking along the bottom.

There are twenty different kinds of armadillos. All of them live from Mexico through most of South America. And one kind is also found in the southern United States: the nine-banded armadillo. Actually, this long-nosed armadillo sometimes has only eight bands between its shoulder and hip plates.

Armadillos come in many sizes. The pink fairy armadillo is the smallest—6 inches (15 cm) from head to tail. The giant armadillo is ten times as long—measuring about 5 feet (152 cm).

Armadillos are timid animals. When frightened, they run for their dens. Inside their burrows they are safe. The pink fairy armadillo often blocks the opening of its den with its blunt, scaly hind end. The pichi (say PEA-chee) armadillo wedges itself into a shallow burrow with the toothlike edges of its jagged scales. It is difficult for an enemy to drag it out.

An armadillo can also protect itself by digging a new place to hide. It rapidly makes a tunnel and vanishes before a predator's eyes. Finally, the three-banded armadillo can roll its body into an armored ball. Such enemies as foxes or wolves cannot get a grip on the smooth plates. Other armadillos curl up only partway to protect their soft bellies.

Most armadillos live alone, but some kinds live in pairs or in small groups. Armadillos may even share a burrow. Most dig their dens in open grasslands. But giant and nine-banded armadillos also live in forest underbrush.

Pichi armadillo: 11 in (28 cm) long; tail, 4 in (10 cm)

△ *Pichi armadillo runs quickly to escape from danger. Once the pichi reaches its grass-lined burrow, the toothlike edges of its armor will help wedge it inside.*

New kind of ball? No, it's a three-banded armadillo. ▷ *When threatened, this animal curls up tightly and hisses. Its head, tail, and feet fit under the plates that cover its curved back. Big cats cannot bite its soft underparts.*

Three-banded armadillo: 9 in (23 cm) long; tail, 3 in (8 cm); diameter rolled up, 4 in (10 cm)

Armadillos make snorting noises as they move about. Hikers near sand dunes in Florida—and in other dry parts of the southern United States—often hear armadillos sniffing for grubs in the tall grass. The animals probably depend on their sense of smell to find food. Some burrow into termite mounds and stay there all day, feeding on insects.

Ants and termites are an armadillo's favorite foods. The animal pushes its worm-shaped tongue far into an insect nest. Its tongue comes out covered with insects. These are quickly gobbled down. Armadillos also eat snakes, worms, snails, beetles, roots, fruit, and sometimes dead animals.

Most armadillos give birth to one offspring or to twins. But the nine-banded armadillo bears four identical offspring. The young in each litter are the same sex—all male or all female.

Pink fairy armadillo: 5 in (13 cm) long; tail, 1 in (3 cm)

△ *Which end points forward? Soft, white hair covers the tiny snout of a pink fairy armadillo. A short fringe flutters from its back end. This kind of armadillo—the smallest of all—weighs only 3 ounces (85 g). A giant armadillo (below) weighs 120 pounds (54 kg)—600 times as much—and measures ten times as long from head to tail. The curving claws on the larger animal's front feet grow as long as the body of the pink fairy armadillo.*

Giant armadillo: 37 in (94 cm) long; tail, 23 in (58 cm)

Ass

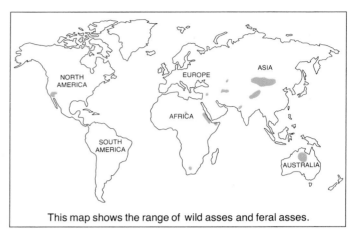

This map shows the range of wild asses and feral asses.

Female asses, called jennies, lead two-month-old foals to water. The young can run just a few hours after birth.

SUREFOOTED AND UNHURRIED, the ass trots along on small, tough, U-shaped hooves. This sturdy animal—a close relative of the horse—is well suited, or adapted, to the harsh, dry lands in which it lives.

Asses can make their way in rocky canyons and up steep hills. They can go without water for several days. They can drink salty water. They can eat leaves, bark, twigs, and thistles, as well as grasses.

For centuries, people around the world have kept asses as pack animals. Tame asses—called burros or donkeys—probably carried supplies for the people who built the pyramids of Egypt thousands of years ago. Columbus brought donkeys with him to the New World. About a hundred years ago, burros carried the belongings of prospectors in the western United States. Some of these burros ran off into the wild. Their descendants are called feral (say FEAR-ul) asses.

The wild ass is smaller than most horses. It has large pointed ears that stick up into the air. A dark stripe usually runs down the animal's back, from its short, wiry mane to its tufted tail. Asses range in color from white to tan to black. The Nubian ass of Egypt has a stripe across its shoulders. Dark bands circle the legs of the Somali ass.

The gray coats of Somali asses blend into their desert environment. A group may travel for days without finding water, eating only tough grasses

ASS

HEIGHT: 38-63 in (97-160 cm) at the shoulder

WEIGHT: 300-1,200 lb (136-544 kg)

HABITAT AND RANGE: dry areas of Africa, the Middle East, Asia, Australia, and parts of North America; domestic asses are found in many parts of the world

FOOD: desert plants

LIFE SPAN: 10 to 25 years in the wild

REPRODUCTION: usually 1 young after a pregnancy of 11 or 12 months, depending on species

ORDER: perissodactyls

Long ears tilted to catch sounds, a male ass, called a ▷ jack, nibbles on thorny shrubs. Asses have adapted well to life in rugged places, like this island off California. ▽ Biting and shoving, two jacks fight for a mate. Battles test strength, but they rarely end in serious injury.

66 North American feral ass: 46 in (117 cm) tall at the shoulder

and small, thorny bushes. To escape such enemies as hyenas, Somali asses climb steep hills and hide among the broken rocks. If attacked, they may bite and kick fiercely to defend themselves. When sandstorms blow, the animals lower their heads and turn their tails to the wind.

The khur (say KUR), another kind of wild ass, lives in Asia. With its shorter ears and broader hooves, this animal looks more like a horse than other asses do. Sand-colored khurs live in dry, open country. These swift animals travel and graze in the cool evening and early morning hours.

Asses are known for their loud braying "hee-haw." The noise they make sounds like a rusty hinge of a door slowly opening. Scientists have found that each ass makes its own special braying sound. One animal can recognize another by its call. Because of their "song," burros in Colorado are sometimes called Rocky Mountain canaries!

Somali ass: 49 in (124 cm) tall at the shoulder

△ *Like striped stockings, black markings cover the legs of a Somali ass feeding on bushes. People in Africa began taming such asses thousands of years ago.*

▽ *Hardy khurs roam a sun-scorched plain in India. In the early morning and in the evening, they travel widely, looking for food. The animals graze on islands of grass that spring up after rains.*

Khur: 46 in (117 cm) tall at the shoulder

Aye-aye

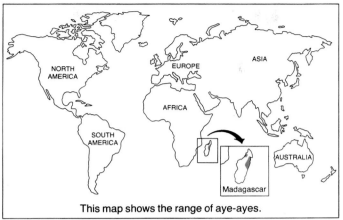

This map shows the range of aye-ayes.

AYE-AYE

LENGTH OF HEAD AND BODY: **14-17 in (36-43 cm); tail, 22-24 in (56-61 cm)**

WEIGHT: **4 lb (2 kg)**

HABITAT AND RANGE: **rain forests of Madagascar**

FOOD: **insect larvae and fruit**

LIFE SPAN: **as long as 5 years in captivity**

REPRODUCTION: **1 young after a pregnancy of unknown length**

ORDER: **primates**

Sensitive ears alert, a male aye-aye (left) makes a nighttime search for food. An aye-aye eats insect larvae. It removes them from under tree bark with its long middle finger (right).

THE ISLAND OF MADAGASCAR, off the coast of Africa, is the only place in the world where the aye-aye lives. This bushy-tailed, big-eared mammal is found in rain forests. It is a member of the primate order, which includes monkeys, apes, and humans.

The aye-aye spends the day sleeping in a ball-shaped nest made of leaves and branches. The animal builds its nest in the fork of a large tree. Each round nest has a hole in the side through which the aye-aye enters and leaves. The nest is a closed, safe place for the animal to rest.

The aye-aye wakes up when the sun goes down. It climbs through the trees searching for food. Sometimes it dangles from a branch by its legs.

With its long, thin fingers, the aye-aye grooms itself. The animal's middle finger is even longer and thinner than the others. It looks like a dry, bent twig. Using this finger, the aye-aye can get insect larvae from under tree bark. The animal listens for the sounds of larvae, then quickly gnaws a hole in the bark with its sharp teeth. It reaches into the hole with its long middle finger and removes the larvae. Aye-ayes also use their middle fingers to scoop out the juice and meat of coconuts.

Female aye-ayes have one offspring at a time. They give birth to tiny young in their nests in the trees. Except for females with offspring, aye-ayes usually live alone.

Babirusa

(say bobby-ROO-suh)

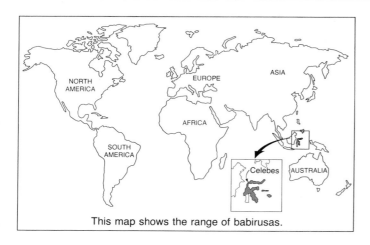

This map shows the range of babirusas.

ON SOME ISLANDS of Indonesia lives a kind of wild hog with curved tusks that grow right through the skin of its snout. Some people there think the tusks look like a deer's antlers. So they call the animal *babirusa*, which means "pig-deer."

Only the male babirusa has tusks. The lower tusks, like those of other wild hogs, are sometimes used for fighting. But its upper tusks are something of a mystery. The babirusa does not use them for self-defense or for rooting because they point in the wrong direction. Sometimes they grow as long as 17 inches (43 cm) and curve around in a circle! Scientists think the upper tusks may help to attract mates.

Dried mud cakes the curling tusks and wrinkled hide of a male babirusa, a wild hog of Indonesia.

△ *After a tropical rain, a male babirusa cools off with a roll in the mud. Like other pigs, babirusas often wallow in mud. They live in dense, swampy forests.*

▽ *Female babirusa towers over her one-month-old young. When danger threatens, she fiercely defends her offspring. Newborn babirusas measure 8 inches (20 cm) long. Within nine months, they grow to their full size.*

The babirusa is a distant relative of the pig that lives on farms. Aside from its curious tusks, a babirusa looks much like a pig. But its legs are longer than a pig's legs. A full-grown babirusa can measure more than 3 feet (91 cm) long and almost 3 feet (91 cm) high. It can weigh as much as 220 pounds (100 kg). Its brownish gray hide looks hairless. The babirusa's skin may be either wrinkled or smooth.

Babirusas have small ears, but their hearing is sharp. This comes in handy, since the animals feed and move around in the dark. It is hard for scientists to study babirusas in the wild because of the animals' nighttime habits.

Babirusas live in moist forests and along the edges of rivers and lakes. Like other members of the pig family, they spend much time wallowing in the mud. Sometimes babirusas swim to nearby islands to feed on water plants, leaves, fallen fruit, and shoots. They also eat insect larvae found in rotting tree trunks.

Before a female babirusa gives birth, she prepares a nest in a hidden place. There she has her young. Usually one or two offspring are born after a pregnancy of about five months. Most other members of the pig family have larger litters. At birth, the babirusas are tiny—only about 8 inches (20 cm) long. Unlike many other young wild hogs, which have striped coats, newborn babirusas have smooth, unmarked skins. You can read more about other kinds of hogs on page 264.

People of the islands where babirusas live have a legend about the animals. They say that when a babirusa wants to sleep, it hangs itself up on a tree branch by its tusks. That way it is out of danger. Actually, the babirusa spends its sleeping hours safely hidden on the ground.

BABIRUSA

LENGTH OF HEAD AND BODY: 35-43 in (89-109 cm); tail, 8-12 in (20-30 cm)

WEIGHT: as much as 220 lb (100 kg)

HABITAT AND RANGE: moist forests and edges of rivers and lakes on Celebes and on nearby islands of Indonesia

FOOD: water plants, leaves, shoots, fruit, and insect larvae

LIFE SPAN: as long as 22 years in captivity

REPRODUCTION: 1 or 2 young after a pregnancy of about 5 months

ORDER: artiodactyls

Baboon

The baboon is a kind of monkey. Read about monkeys on page 376.

Badger

(say BADGE-er)

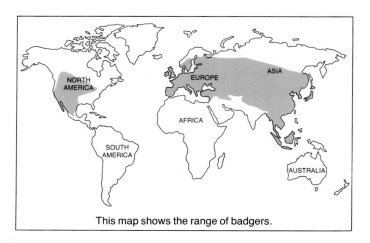

This map shows the range of badgers.

BADGER

LENGTH OF HEAD AND BODY: 13-31 in (33-79 cm); tail, 4-7 in (10-18 cm)

WEIGHT: 4-37 lb (2-17 kg)

HABITAT AND RANGE: dry open plains, woodlands, mountains, tropical forests, and prairies of Europe, Asia, and parts of North America

FOOD: rodents, small mammals, birds, snakes, frogs, insects and their larvae, worms, fruit, and roots

LIFE SPAN: 11 to 13 years in captivity, depending on species

REPRODUCTION: 1 to 4 young after a pregnancy of 2 to 10 months, depending on species

ORDER: carnivores

WHETHER IT IS LOOKING FOR FOOD or for a safe place to sleep, the badger usually digs a hole. This strong, short-legged animal loosens the dirt with the long claws at the ends of its forefeet.

In North America, badgers usually live alone on dry plains and on open prairies. There they can easily dig holes to find their prey: mice, pocket gophers, ground squirrels, and insects. During warm months, some badgers dig a new hole every day. There they rest and escape from the hot sun. In very cold weather, they dig a burrow and remain inside for several months. They sleep most of the time, waking only occasionally to look for food.

In Europe and Asia, badgers live in woodlands and meadows, on mountains and plains. The animals usually dig in dry, loose soil. European badgers live in groups called clans. Together they make setts—underground networks of rooms and tunnels. Some setts are hundreds of years old. Many generations of badgers may have enlarged the same sett. European badgers eat more plants and worms

Poking up through the sand, a North American ▷
badger emerges from a new entrance to an old burrow.
Badgers often snort and snuffle as they dig.

Kicking up the dirt, a North American badger (below, left) tunnels under a rocky cliff. Another pursues a mouse it chased out from underground. Badgers dig for food and shelter with their long, curved claws.

North American badger: 29 in (74 cm) long; tail, 5 in (13 cm)

than their American relatives do. They eat earthworms, sucking them in like spaghetti. They feed on fruit, plants, and small animals.

Litters of one to four young are born inside a burrow. A European badger has its offspring in a nest inside its sett. To make its nest, the badger clutches leaves and grass between its chin and forelegs. It shuffles backward into its sett. Most young badgers drink their mother's milk for four or five months. Then they begin to find food for themselves.

European badgers are playful. Both adults and young—called cubs—chase and tumble and even play a game that looks like King of the Mountain. These games are important in a badger's life. Cubs learn to defend themselves. And games make the ties between badgers stronger.

Badgers become fierce, even vicious, when cornered by such enemies as dogs and foxes. Facing attackers, badgers bristle their fur and look larger. Their skin is so tough that it is difficult for enemies to bite into them. With powerful jaws, strong teeth, and sharp claws, badgers are savage fighters.

Because European badgers live in groups, they

European badger: 29 in (74 cm) long; tail, 6 in (15 cm)

communicate with each other by various sounds and smells. A growl means a badger may attack. A mother calls her young with a high-pitched cry.

Badgers are relatives of skunks, weasels, martens, and polecats. All these animals have scent glands. By making a scent mark along its path, one badger can let another badger know it has passed by. Scent marks also show the way to feeding grounds. The stink badger of southeastern Asia can spray its enemies, just as a skunk does.

Read about another badger relative, the ratel, or honey badger, on page 487.

△ Following a scent, a ferret badger sniffs out its prey. Hunting at dusk or during the night, this Asian badger eats worms, insects, rodents, and fruit. It can climb trees to find small birds and eggs. During the day, it may sleep in a rocky shelter or occasionally on the branch of a tree.

◁ European badger trots across the countryside to the safety of its burrow. It can move surprisingly fast on its short legs. These badgers travel well-worn paths from their setts, or burrows, to feeding grounds. They mark the way with scent.

▽ Resting on leaves, a hog badger of Asia finds roots and worms with its long, blunt snout. Hog badgers often hide in deep burrows during the day. Strong teeth and sharp claws help these animals defend themselves.

Hog badger: 24 in (61 cm) long; tail, 7 in (18 cm)

Bandicoot

Rabbit-eared bandicoot: 15 in (38 cm) long; tail, 8 in (20 cm)

Barred bandicoot: 12 in (30 cm) long; tail, 4 in (10 cm)

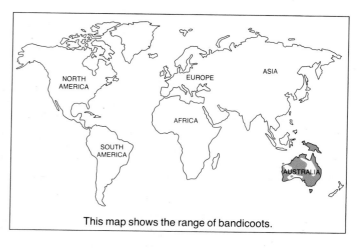

△ *Barred bandicoot sniffs for insects in the soil. The pouch of this striped marsupial opens to the rear.*

◁ *Grains of sand cling to the face of a rabbit-eared bandicoot—or bilby—as it digs an underground home in a dry desert in Australia. The tunnels of its burrow can extend downward as far as 5 feet (152 cm).*

WHEN NIGHT FALLS in Australia, New Guinea, and neighboring islands, small sharp-nosed animals scurry from their nests and shallow burrows. These frisky animals are called bandicoots. After dark, they scamper about searching for food. With the sharp claws on their front feet, the animals scratch in the soil for insects and worms. Some kinds of bandicoots also feed on young mice and plants.

Bandicoots usually hunt and live alone. Most kinds make shallow burrows or grassy nests. These homes help protect them from their enemies. They also shelter bandicoots from the heat of the desert areas where the animals often live. The rabbit-eared bandicoot digs its den deeper than other kinds—as much as 5 feet (152 cm) underground.

Bandicoots, like koalas and kangaroos, belong to the group of pouched mammals called marsupials (say mar-soo-pea-ulz). There are 19 kinds of bandicoots.

Just two weeks after mating, a female gives birth to as many as six tiny, underdeveloped young. Bandicoots spend their first two months in the safety of their mother's pouch. A bandicoot's pouch is different from a kangaroo's pouch. It opens to the rear! Inside the pouch, young bandicoots are shielded from flying dirt when their mother digs for food or shelter.

BANDICOOT

LENGTH OF HEAD AND BODY: 7-22 in (18-56 cm); tail, 4-10 in (10-25 cm)

WEIGHT: about 2 lb (1 kg)

HABITAT AND RANGE: plains, deserts, and forests of Australia, New Guinea, and neighboring islands

FOOD: mostly insects, but also lizards, mice, snails, worms, and some plants

LIFE SPAN: 3 to 7 years in captivity, depending on species

REPRODUCTION: 1 to 6 young after a pregnancy of about 2 weeks

ORDER: marsupials

This map shows the range of bandicoots.

Barbary ape

The Barbary ape is a kind of monkey. Read about monkeys on page 376.

Bat

BAT

LENGTH OF HEAD AND BODY: 1-16 in (3-41 cm); tail, about 1-3 in (3-8 cm); wingspan, 6 in-6 ft (15-183 cm)

WEIGHT: less than $\frac{1}{4}$ oz-2 lb (7 g-1 kg)

HABITAT AND RANGE: all kinds of habitats worldwide, except in the Antarctic and in parts of the Arctic

FOOD: insects, fruit, nectar, pollen, flowers, small animals, fish, and blood

LIFE SPAN: as long as 30 years in the wild, depending on species

REPRODUCTION: usually 1 young after a pregnancy of $1\frac{1}{2}$ to 8 months, depending on species

ORDER: chiropterans

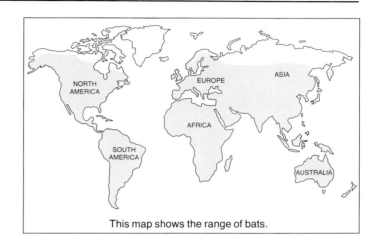

This map shows the range of bats.

IN GHOST STORIES, no haunted house is complete without a few bats flying around. And Halloween brings to mind scary images of witches, goblins, and bats. Bats seem frightening and mysterious. They dart around at night, hang upside down in caves, and roost in abandoned buildings. But most stories about bats aren't true. Bats don't get tangled in people's hair. And they are not blind.

All bats can see, but many do not use their eyes to find food. Instead, they use their ears. Even in the dark, a bat can find its way, can avoid obstacles, and can detect food by using echolocation (say ek-oh-low-KAY-shun). When flying, a bat sends out a series of short, high-pitched beeping sounds through its mouth or its nose. It listens for the echoes that bounce back when *(Continued on page 80)*

Awakening at night, two bats circle at the entrance of the home they share in a hollow tree.

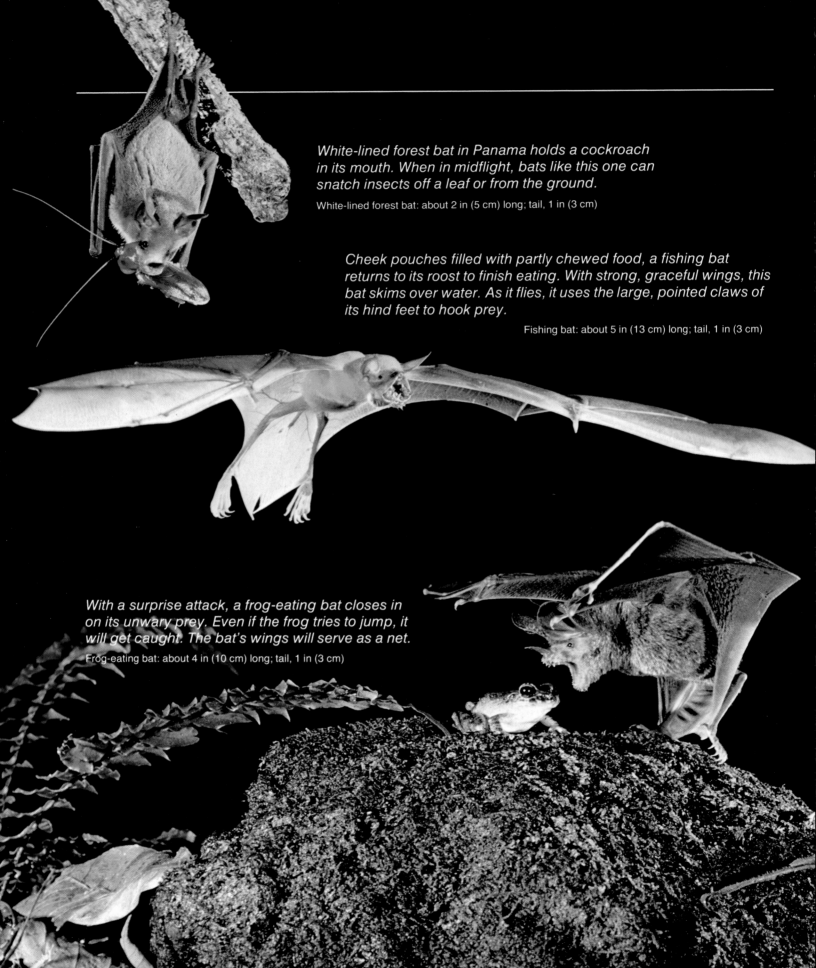

White-lined forest bat in Panama holds a cockroach in its mouth. When in midflight, bats like this one can snatch insects off a leaf or from the ground.

White-lined forest bat: about 2 in (5 cm) long; tail, 1 in (3 cm)

Cheek pouches filled with partly chewed food, a fishing bat returns to its roost to finish eating. With strong, graceful wings, this bat skims over water. As it flies, it uses the large, pointed claws of its hind feet to hook prey.

Fishing bat: about 5 in (13 cm) long; tail, 1 in (3 cm)

With a surprise attack, a frog-eating bat closes in on its unwary prey. Even if the frog tries to jump, it will get caught. The bat's wings will serve as a net.

Frog-eating bat: about 4 in (10 cm) long; tail, 1 in (3 cm)

Hummingbird of the night, a lesser long-tongued bat sips nectar from a tropical flower in Panama. Using its short, broad wings, the bat can hover at a flower while it feeds. This bat also eats fruit and insects.

Lesser long-tongued bat: about 3 in (8 cm) long; tail, 1 in (3 cm)

Sharp claws on its toes give a tube-nosed bat in Australia a good grip on a branch. This bat spends its day roosting upside down, wrapped in its wings.

Tiny suction cups on wrists and ankles help two sucker-footed bats walk on a smooth surface. These small bats live in partly opened leaves.

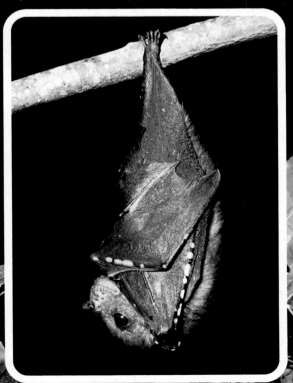

Tube-nosed bat: about 4 in (10 cm) long; tail, 1 in (3 cm)

Sucker-footed bat: about 2 in (5 cm) long; tail, 1 in (3 cm)

the beeps hit an object. From these echoes, a bat can tell where an object is and whether it is moving. The tones of the beeping sounds that each kind of bat produces are different. People don't notice the sounds that most bats make because they are beyond the range of human hearing.

Bats are the only mammals that can fly. Flying lemurs and flying squirrels can move through the air, but they really only glide. Bats flap their wings and fly. Two delicate layers of skin stretch from the sides of a bat's body to the ends of its long finger bones. The wings are moved by powerful muscles that help the animal fly easily through the air. Some bats can fly 40 miles (64 km) an hour or more.

There are more than 900 kinds of bats—with wingspans that range from 6 inches (15 cm) to 6 feet (183 cm). Some bats have tails, and some do not. Scientists divide the animals into two main groups. In one group are the microchiropterans (say my-crow-kye-ROP-tuh-runs). Bats in this group tend to be small. They have large ears and small eyes. Microchiropterans use echolocation. In the other group are the megachiropterans (say meg-uh-kye-ROP-

tuh-runs). These bats are usually larger. They have small ears and large eyes. Most megachiropterans do not use echolocation.

Most bats are microchiropterans. These bats feed mainly on insects and other small animals. Some microchiropterans also feed on fruit and flowers. They live everywhere in the world, except in the Antarctic and in parts of the Arctic.

Some bats that live in colder areas may fly to warmer places before winter arrives and food becomes hard to find. Others sleep for months at a time. This kind of sleep is called hibernation (say hye-bur-NAY-shun). When a bat hibernates, its body temperature drops. Its heart rate and breathing slow down. And it lives off fat stored in its body.

Most megachiropterans live in Africa, in Asia, and in Australia. They feed mainly on fruit and flowers. Most fruit- and flower-feeding bats—in either group—are very much alike. They use their keen sense of smell to detect flowers and ripe fruit. To lap nectar from inside a blossom, a flower-feeding bat uses its pointed snout and a tongue that may measure one-third the length of its body.

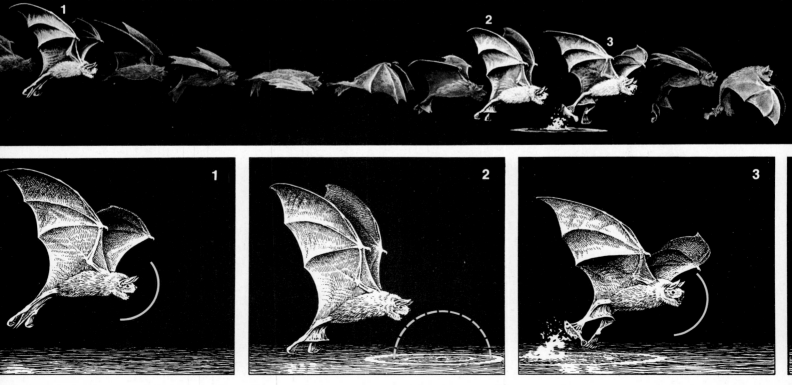

Two ways to look at a fishing bat catching prey: The top drawing shows the animal in motion. The bottom

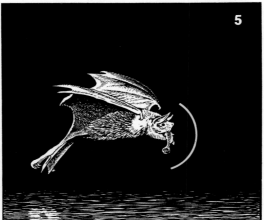

△ Wings flutter and flap in a cave in Trinidad. Even in such crowded conditions, bats rarely collide. Bats usually begin to stir toward evening. As they wake up, they circle the cave in orderly flights. They then leave the cave to feed.

◁ Using echolocation to search for food and to find its way, a fishing bat approaches the water (1), beeping high-pitched sounds (each shown as a solid line). As a guppy's head breaks the surface (2), it reflects a sound, causing an echo (shown as a broken line). Beeping again, the bat hooks the fish with its curved claws (3). As it brings the prey to its mouth (4), the bat remains silent. Chewing on the fish in midair (5), the bat sends out another signal. The bat actually would produce about 14 beeps during this hunt, although the drawing shows only 3 of them. From start to finish, the action takes just half a second!

drawing freezes the movement at certain points in the hunt.

Vampire bat: about 4 in (10 cm) long

Blossom bat: about 3 in (8 cm) long

△ *Razor-sharp teeth of a vampire bat (above, left) in Trinidad make a painless cut on a donkey's foot to get at blood. After a meal of nectar and pollen, a blossom bat (above, right) in Australia grooms one of its wings.*

◁ *Attic of a house in France provides a roost for a group of long-eared bats. Huge ears help the animals find insects by echolocation.*

Long-eared bat: about 2 in (5 cm) long; tail, 2 in (5 cm)

Insect eaters use echolocation to find their prey. These bats feast on insects that fly through the night air. A bat may catch a big insect with its mouth. It may scoop a small insect up with a wing and pull its victim to its mouth. Some bats can gobble 12 or more mosquito-size insects in a minute.

Fishing bats also hunt by echolocation. They use their huge claws to catch fish. Flying low over water, these bats can detect a fish breaking the surface. Then they reach down and hook their catch with their sharp, curved claws.

A vampire bat has razor-sharp teeth that it uses to make a shallow, painless cut in the skin of its prey. With rapid movements of its tongue, it laps up blood from the cut. This small bat is only about 4 inches (10 cm) long—about the size of a mouse. But it can drink its weight in blood—about 1 ounce (28 g)—each night. A bat may drink so much that it becomes too full to fly! Vampire bats rarely bite people. They often feed on the blood of chickens, cattle, donkeys, and deer. The wounds they make can become infected. And vampire bats sometimes carry a disease called rabies.

Most bats have only one offspring a year. Many kinds give birth in nurseries where large numbers of bats are found. A nursery may contain fewer than a hundred or as many as several million females and their young. A newborn bat clings to its mother or to the ceiling of the nursery. Within two to twelve weeks, the young bats will begin to fly.

Bats benefit people in several ways. Some feed on harmful insects. Others pollinate flowers as they fly from blossom to blossom sipping nectar. Seeds dropped by fruit bats may sprout into plants.

Scientists are studying echolocation. They would like to perfect a similar system that would allow blind people to detect objects with sound.

◁ *Two hoary bat offspring cling to the belly of their mother, hanging upside down in a tree. The hoary bat usually gives birth to twins—or sometimes even triplets.*

Hoary bat: about 3 in (8 cm) long; tail, 2 in (5 cm)

Pink and naked, thousands of young bent-winged bats squeak and squirm in a cave in Australia. Huddling together keeps these young bats warm. Two mothers (with fur) have returned from hunting to nurse their young. After about three weeks, the young bent-winged bats will have fur and will begin to practice flying.

Bent-winged bat: about 2 in (5 cm) long; tail, 2 in (5 cm)

Wrapped in its wings, a greater horseshoe bat (below) in France wakes up from hibernation.

With specially shaped noses and ears, the bats below get food and find their way. Large eyes and a sensitive nose help a bat called a black flying fox (below, right) find food. The large ears of a yellow-winged bat (below, center) pick up sounds from its prey. Flaps on the snout of a hammer-headed bat may help it call and attract mates.

Greater horseshoe bat: about 3 in (8 cm) long; tail, 1 in (3 cm)

Hammer-headed bat: about 11 in (28 cm) long

Yellow-winged bat: about 3 in (8 cm) long

Black flying fox: about 10 in (25 cm) long

Bear

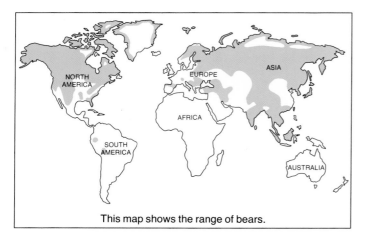

This map shows the range of bears.

"GRIZZLY!" JUST THE WORD could frighten settlers in the Old West. Pioneers often lived in wild regions where grizzly bears roamed. And they knew that a disturbed or wounded grizzly could kill a person with one swipe of its paw. American Indians also feared and respected the grizzly. As a test of bravery, young men of some tribes had to kill a bear using only a bow and arrows. Its claws were strung into a necklace and worn with great pride.

Grizzlies are a type of brown bear—the most wide-ranging of the seven different kinds of bears. The animals are called grizzlies because their thick brown fur is tipped with lighter-colored hairs. The

animals' coats look grizzled, or streaked with gray.

Once, grizzly bears lived throughout western North America. But now they are found mainly in mountainous areas of Wyoming, Montana, Alaska, and western Canada. Grizzlies grow very large. Males reach a length of about 8 feet (244 cm) from head to rump and weigh about 800 pounds (363 kg).

Another, much larger, brown bear is found on the Alaskan coast. These giant bears measure as long as 10 feet (305 cm) and weigh as much as 1,700 pounds (771 kg). Alaskan brown bears and polar bears are the largest meat-eating land mammals in North America.

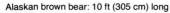
Alaskan brown bear: 10 ft (305 cm) long

Grizzly bear: 8 ft (244 cm) long

△ *Grizzly bear in Alaska stands up in a meadow and takes a look around. Grizzlies have a keen sense of smell. As they search for food, they stop often to sniff the air.*

◁ *After a nap, a female Alaskan brown bear and her cub head for a river to catch fish.*

◁ Grizzlies feed on wild berries. In the summer and fall, these brown bears eat large amounts of rich food and grow fat. The fat will nourish them through the winter.

Bears are very adaptable animals: That means they can fit into many kinds of environments. Bears will eat whatever food is available or in season. Though some bears are huge animals, they feed mainly on plants, fruit, and insects. Some use their long, curved claws to dig for roots. Bears also eat small mammals and the remains of dead animals.

Bears are loners and usually wander by themselves in search of food. Usually only females with cubs feed together. But in summer, when salmon swim up coastal rivers to lay eggs, as many as eighty Alaskan brown bears may gather together to fish. A bear waits on shore until it spots a salmon. Then, with a leap, it belly flops into the water and pins the fish to the bottom with its paws or mouth. A bear may catch six to eight salmon before it has eaten enough and lumbers off to rest.

In summer and autumn, brown bears—like other bears that live in cold or moderate climates—

BEAR

LENGTH OF HEAD AND BODY: 43 in-10 ft (109-305 cm)

WEIGHT: 55-1,700 lb (25-771 kg)

HABITAT AND RANGE: mountains, forests, swamps, and grassy plains in parts of North and South America, Europe, and Asia

FOOD: grasses, roots, berries, insects, fruit, eggs, birds, fish, and other animals and their remains

REPRODUCTION: usually 1 to 4 young after a pregnancy of 7 to 9 months

LIFE SPAN: as long as 35 years in the wild, depending on species

ORDER: carnivores

▽ *In late summer, Alaskan brown bears gather at a river to fish for salmon swimming upstream to lay eggs. The biggest males take the best fishing spots. Mothers with cubs usually take the next best. In the small picture, a hungry bear eats its meal shoulder-deep in water. Some bears carry their catch to land.*

eat large amounts of rich food and grow very fat. As the weather turns colder, their fur becomes thick. They start to prepare dens where they will sleep during the winter, when food is hard to find. Brown bears dig their dens in hillsides. Black bears may find places in caves, under dead trees, or anywhere they will be sheltered from bad weather. They bring in leaves, branches, and grasses to line the den. Some bears return to the same area every year.

When winter arrives, bears retreat to their dens. Their fat nourishes them throughout the cold months, and their heavy coats keep them warm. Bears sleep through the winter. Their sleep is a kind of hibernation (say hye-bur-NAY-shun), but their body temperatures do not drop sharply. Bears can be awakened easily during the winter. They may even leave their dens for short periods of time.

In North America and in Europe, a female bear—called a she-bear—gives birth to young in her den. A litter of two or three bear cubs is usually born in midwinter. The blind, helpless cubs weigh about a pound (454 g) at birth. Their mother may nurse them for several months. If little food is available, she may nurse them much longer.

In the spring, bears come out of their dens and start looking for food. The adults may weigh much

△ *American black bear looks out from a tree branch. Good climbers, black bears live mainly in forests.*
▽ *Black bear lumbers through a meadow in Montana. The most common bear in North America, the black bear lives from Canada through northern Mexico.*

less than when they entered their dens. The cubs now have soft, fluffy fur. They are playful and curious, but they stay close to their mother. If a she-bear thinks her cubs are in danger, she will move quickly to protect them. She will charge an animal that threatens her young.

Bear cubs stay with their mother for about two years. They follow her and learn to search for food and to defend themselves. The cubs spend the next winter with their mother in the den. Then, when they are old enough to care for themselves, they wander off on their own.

A young bear would have little chance of surviving without its mother's protection. But one orphaned black bear cub became famous. Most people know him as Smokey Bear. He was rescued after a

◁ *Black bear nibbles on a branch. Except for females with cubs, bears usually live and hunt alone. They eat almost anything—insects, berries, roots, small mammals, and dead animals they find.*

▽ *Born in a winter den, black bear cubs only a few months old begin to explore the world outside with their mother. Newborn cubs weigh less than a pound (454 g). They nurse for a few months, or perhaps longer. Cubs stay with their mother for about two years.*

American black bear: 5 ft (152 cm) long

Polar bear: 10 ft (305 cm) long

△ *Powerful swimmer, a polar bear paddles with its front paws and steers by moving its back legs. People have seen polar bears hundreds of miles from land. The animals float much of that distance on ice. But they swim strongly in search of prey—usually seals.*

Something moves against the arctic landscape! Coal black eyes and nose mark an approaching ▷ polar bear. Its white coat blends well with the snowy terrain.

forest fire when he was only a few months old. A game warden named him Smokey and cared for the badly burned cub. Smokey recovered and became a living national symbol for fire prevention.

Black bears are the most common bears in North America. They live mainly in the mountains, forests, and swamps of the United States, Canada, and northern Mexico. They often wander through the woods, breaking off branches and eating acorns or berries. Skillful climbers, black bears can quickly go up tree trunks by gripping the bark with sharp, curved claws. Some even make winter dens in holes in trees, as high as 60 feet (18 m) above the ground.

Although the animals are called black bears, the color of their fur can range from black to brown to a rarer blue-gray or white. These medium-size bears weigh about 300 pounds (136 kg) and reach about 5 feet (152 cm) in length.

Many black bears live in national parks, and they sometimes become tourist attractions. Although visitors are warned not to feed the bears, they often do. These bears may seem tame, but if one of them is irritated—watch out! All bears can be dangerous if angered or bothered. When they

are left alone, bears usually will not harm anyone.

Some black bears are known as campsite thieves because they are always searching for something to eat. Bears love honey. Finding a bee tree, a bear will sit down and eat its fill, despite the stings of angry bees. But its sweet tooth can lead to a

▽ *Curled into a cradle, a female polar bear in Alaska naps and shelters her cub. Dense fur and a thick layer of fat protect polar bears from the icy temperatures of their arctic home.*

Short, sleek fur covers the body of a Malayan sun ▷ *bear. Smallest of all bears, these animals get their name from the light fur on their chests. They spend some of their time in trees, sleeping by day and feeding by night.*

▽ *V-shaped marking looks like a collar around the neck of an Asiatic black bear, or moon bear. These animals live in mountain forests. Like all bears in Asia, they feed mainly on plants, insects, and fruit.*

Asiatic black bear: 4 ft (122 cm) long

Malayan sun bear: 43 in (109 cm) long

problem: The bear is one of the few mammals in the wild that can get cavities!

Bears look awkward and slow moving. Their bodies are bulky, and they shuffle along on flat feet. But don't be fooled! Bears can run quickly to chase prey or to escape danger.

The polar bear lives in the icy wilds of the Arctic. Polar bears prey mainly on seals. Floating on pack ice, they travel great distances in search of food. Polar bears have been spotted hundreds of miles from land as they hunt seals. These huge bears take naturally to the water. The toes on their front paws are slightly webbed. And they have broad feet for paddling. By moving their back legs, they can control the direction they are swimming.

Polar bears are well adapted, or suited, to their icy home. A thick layer of fat and a coat of dense fur

keep them warm. Fur on the bottom of their paws prevents them from slipping on ice. Their white coats help them blend into the snowy landscape. Males grow as long as 10 feet (305 cm) and weigh as much as 1,700 pounds (771 kg).

Polar bears build their dens in snowbanks. They tunnel down into the drifts and dig out a small room. A female usually gives birth to twin cubs in these winter dens. The male bears will continue hunting throughout the winter.

Bears that live in warmer climates normally stay active most of the year. The Malayan sun bear lives in tropical forests in Southeast Asia. It spends some of the day resting in a nestlike bed of branches in a tree. Plants and insects are its main foods.

The sloth bear of India and Sri Lanka also feeds on insects and plants. With its long, curved claws, it

European brown bear: 4 ft (122 cm) long

△ *European brown bear climbs a tree in its forest home. People sometimes capture these bears and train them to perform in circuses.*

tears open termite mounds. Then it sucks up the insects, using its lips and long, flexible snout.

The Asiatic black bear roams mountain forests. It sometimes sleeps during the day in caves. At night, the animal climbs trees in search of nuts, fruit, and honey. Because the V-shaped marking on its chest looks a little like a crescent moon, it is sometimes called a moon bear.

The rarely seen spectacled bear of South America gets its name from the whitish circles around its eyes. It looks like it is wearing glasses! In its mountainous home, it climbs trees to find fruit to eat. It often sleeps high among the moss-covered branches.

Spectacled bear: 5 ft (152 cm) long

△ *Young spectacled bear stays almost out of sight high in the trees (top) in South America. In a closeup view (above), the bear hangs from a branch and looks for figs and other fruit to eat. Few people have seen these animals—the only bears in South America. They roam parts of the Andes. As the mountain forests in which they live disappear, the bears may become even rarer.*

△ Home for perhaps a dozen beavers, a lodge built of sticks, mud, and stones rises from a mountain pond in Wyoming. The lodge shelters the beavers and protects them from such enemies as bears, wolves, and coyotes. Holding forepaws against its chest, a beaver darts toward an underwater ▷ entrance to its lodge. Webbed hind feet work like swim fins. The broad tail acts as a rudder. It helps the beaver steer left or right, up or down.

94

Beaver

North American beaver: 3 ft (91 cm) long; tail, 12 in (30 cm)

BEAVER

LENGTH OF HEAD AND BODY: about 3 ft (91 cm); tail, about 12 in (30 cm)

WEIGHT: 30-70 lb (14-32 kg)

HABITAT AND RANGE: rivers, lakes, and streams near woodlands in North America and in parts of Europe and Asia

FOOD: bark, twigs, leaves, roots, and aquatic plants

LIFE SPAN: 10 to 12 years in the wild

REPRODUCTION: usually 2 to 4 young after a pregnancy of 3 or 4 months

ORDER: rodents

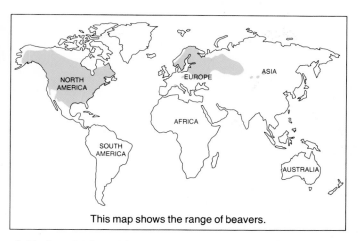

This map shows the range of beavers.

◁ *Sleek, whiskered beaver emerges from the water to repair its lodge. With its forepaws, the animal can hold sticks and stones, roll logs, and scoop up mud.*

THE BUSY BEAVER is one of the few animals—other than man—that can alter its environment. Sometimes beavers dig burrows in the banks of lakes and large rivers. But where the water is not deep enough for their way of life, beavers change the landscape. They build dams across streams. Soon the water deepens behind the dams, turning fields and forests into a watery world of beaver ponds.

On land, the beaver moves awkwardly. The bulky animal—about 3 feet (91 cm) long—cannot escape easily from bears and wolves as it waddles

through the woods. In water, however, the beaver is a strong and graceful swimmer. It often builds its home in the middle of a pond. In this island home, or lodge, the beaver family usually is safe.

A beaver is well suited for life in the water. Long hairs—called guard hairs—protect its thick underfur. This heavy coat helps a beaver stay warm, even in icy water. With a special split claw on each hind foot, a beaver can comb the fur and keep it clean. The beaver combs in oil produced by glands in its body. This helps to make its coat sleek and waterproof.

The animal's hind feet serve as swim fins. These large webbed feet can push a beaver through the water at speeds of 5 miles (8 km) an hour. Its broad, scaly tail acts as a rudder. The beaver uses it to steer right or left, up or down. Sometimes a beaver signals with its tail, too. It slaps the water, warning other beavers when danger threatens.

A beaver can stay underwater for as long as 15 minutes without coming up for air. It holds its breath, and its nose and ears shut tightly when it dives. A beaver has a set of transparent eyelids that slide across its eyes. These serve as goggles, protecting the animal's eyes underwater.

A beaver can close its mouth by pressing together flaps of skin behind its front teeth. The beaver can then chew on wood underwater without getting wood or water down its throat.

Beavers live throughout forested areas in North America and in parts of Europe and Asia. There they can find the trees they need for food and for building lodges and dams. As a beaver cuts down a tree, it usually stands on its hind legs and leans back on its tail. Tilting its head to one side, the animal bites into the tree with its sharp teeth. Chipping deeper and deeper, it cuts through the trunk. As the tree falls, the beaver scampers out of the way.

The beaver then trims off the branches and bark

Inside their lodge, young beavers—called ▷ *kits—greet their mother with high-pitched cries. The kits stay with the family for two years, even after the birth of another litter.*

▽ *Holding its dinner in its paws, a beaver nibbles on a young, tender plant. The kit at its side chews on a leaf. In late summer and early fall, beavers anchor branches near an underwater entrance to their lodge. They use these for food during the winter.*

Kits cling to a parent on one of their first outings. The ▷
young float easily in the water soon after birth. As they
grow heavier, the kits learn to swim and to dive.

and cuts the trunk into smaller pieces. It takes most of the wood to a stream by dragging it or by floating it down canals it has dug. It uses some of the logs to build a dam. With stones and mud, the beaver anchors the logs and branches to the muddy bottom of a stream. It wedges sticks in between the logs. More stones and brush go on top.

As the dam gets bigger, a deep pond forms behind it. The pond must be deep enough so that the water at the bottom does not freeze, even in the coldest weather. In the winter, under a layer of ice, a beaver still continues its underwater life. It swims into and out of its lodge and brings in pieces of branches it keeps near the entrance of its home.

Beavers often build lodges near the middle of their ponds. They anchor sticks in the muddy bottom. They pile stones and more wood on top until the dome-shaped lodge shows above water. They gnaw underwater openings into the lodge and hollow out living quarters, just above the water level. A platform provides a place for drying off and for eating. Shredded wood covers the sleeping area.

▽ *Underwater tunnels lead to the living quarters of a beaver lodge. This drawing shows how a lodge looks on the inside. On a platform, a beaver munches bark on a stick. In this dry area, beavers also sleep and raise their young. Outside, another beaver repairs the lodge with more sticks and mud.*

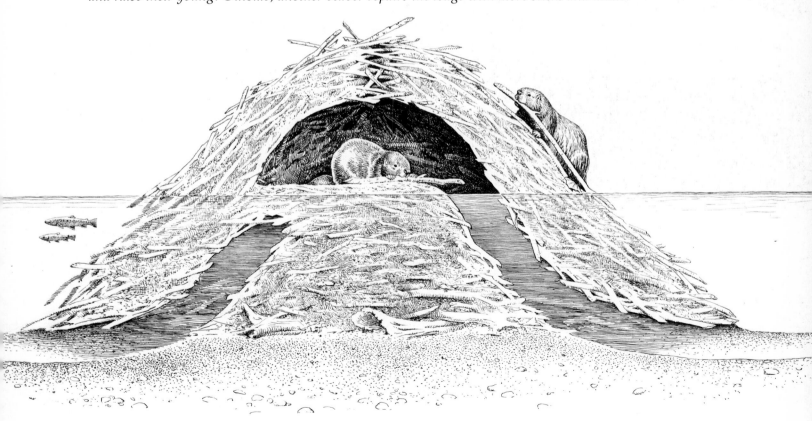

Beavers seal their lodge with mud. They fill almost all the cracks so that winter wind and rain cannot get in. But beavers always leave vents in the roof. On frosty days, you might see the warm air from inside the lodge rising through the vents.

A beaver family lives together in the lodge: the mother and father, the young of the year before, and the recent litter. Usually from two to four young—called kits—are born each spring or summer.

◁ *Tim-ber! A beaver cuts down a small tree by gnawing its way around the trunk. It uses logs of willow, poplar, and alder to build a dam or a lodge.*

▽ *Gigantic beaver dam dwarfs a woman in Alaska. Some beaver dams rise 12 feet (4 m) high and stretch longer than a football field. It takes several generations of beavers to build and maintain them.*

Binturong

Stiff white whiskers curve out from the face of a binturong, a rarely seen mammal of southern and southeastern Asia.

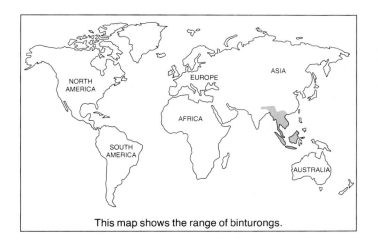

This map shows the range of binturongs.

BINTURONG

LENGTH OF HEAD AND BODY: 24-38 in (61-97 cm); tail, 20-33 in (51-84 cm)

WEIGHT: 20-44 lb (9-20 kg)

HABITAT AND RANGE: dense forests of southern and southeastern Asia

FOOD: fruit and small animals

LIFE SPAN: more than 20 years in captivity

REPRODUCTION: 1 to 4 young after a pregnancy of 3 months

ORDER: carnivores

FROM ITS TUFTED EARS to the end of its muscular tail, a binturong is covered with thick fur. The black hairs of the animal's coat are often tipped with white or dark red. The binturong uses its strong, heavy tail almost like an extra hand as it moves through the trees. The tail often grips a branch tightly as the animal reaches for food with its front feet. Young binturongs can even hang by their tails.

During the day, binturongs rest. But at night they look for food. Their keen sense of smell leads them to fruit and small animals.

The female binturong bears from one to four offspring. After about seven weeks, the young begin to climb out of their nest and explore. They will reach their full size—about 5 feet (152 cm) from head to tail—in about a year and a half.

Binturongs communicate with each other by scent. Each animal has a special gland under its tail that produces a strong-smelling oil. As a binturong travels, it sometimes stops and rubs its hindquarters against a branch. This makes a scent mark and lets other binturongs know it has passed by. The binturong is a kind of civet. All civets communicate with scent. Read about other civets on page 154.

99

Wading through fresh snow, a herd of bison looks for food in Wyoming's Grand Teton National Park.
To reach the grass, bison push snow aside with their muzzles. Bison once roamed most of North America.
Later, settlers hunted them nearly to extinction. Bison now live on preserves and on ranches.

Bison

(say BICE-un)

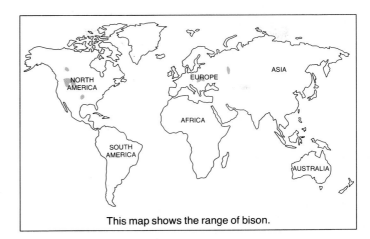

This map shows the range of bison.

DID YOU KNOW that Buffalo Bill never shot a single buffalo? Or that the "home where the buffalo roam" is not part of the United States? The large, shaggy animals pictured on coins and described in folklore are really bison—though the animals are sometimes called American buffaloes.

In the past, many herds of bison roamed North America from Oregon to New York State and from Canada to Mexico. Today these animals live mainly in parks and on wildlife preserves on the Great

Bison: 6 ft (183 cm) tall at the shoulder

Light snow dusts the shaggy winter coat of a bison ▷ *feeding on prairie grasses in South Dakota. Its sharp, curved horns may grow more than 2 feet (61 cm) long.*

Plains and in the Rocky Mountains. Their buffalo relatives are found in Asia and in Africa. Find out more about buffaloes on page 106.

The bison—the heaviest land animal in North America—can weigh a ton (907 kg) or more. An adult male, called a bull, measures about 6 feet (183 cm) tall at the shoulder. A female, known as a cow, is smaller. Both males and females have short, curved horns and large humps on their shoulders. In spite of their size and weight, bison can move surprisingly fast. When danger threatens, the animals can run at 30 miles (48 km) an hour.

In winter, matted, woolly hair covers most of a bison's body. A longer, darker mane hangs from the animal's head, neck, and shoulders. This heavy hair protects the animal. A bison begins to shed its

winter coat as the weather gets warmer. It rubs its body against a tree, a rock, or even another bison to help remove the hair. The bison stops this scratching by summer, when its outer coat has dropped off. In warm weather, insects often bother the bison. It gets rid of the pests by rolling in the dust.

Bison usually graze in the morning and in the evening and rest during the day. Like many hoofed animals, bison do not chew their food fully before swallowing. Later, when resting, they bring up wads of food, called cuds. After chewing these cuds thoroughly, they swallow them and digest the food.

For most of the year, cows and adult bulls live separately. With the approach of the mating season in the summer, they come together and form large herds. Restless males begin to grunt and paw the

Bison plunge into a river in Montana to drink and to cool off. The calves at far left and at far right wear only stubby horns. Their small humps will grow to full size in a few years.

102

Male bison rolls in dust before challenging a rival. Bulls also wallow to rid their coats of pests.

Young calf nurses on a prairie in Nebraska. Darker hair will grow in to replace its reddish brown coat.

ground. Ramming their heads together, two bulls fight to see which is stronger. But rivals rarely battle to the death. Usually the weaker bull signals surrender by turning its head to one side.

In spring, after a nine-month pregnancy, a cow bears one young. The calf, reddish in color, is born without horns or a hump. Within two months, horns sprout and a hump begins to form. Gradually, dark brown hair grows in.

At one time, more than fifty million bison lived in North America. The Great Plains held the largest number. Indians there depended on the bison for survival. They ate bison meat and used the hides to make clothes, tepees, and canoes. Bison also were important in the religions of these Indians.

Settlers who moved west, however, viewed the bison as a nuisance. Shooting the animals became a favorite sport on the frontier. And traders killed the

bison for their hides and tongues. By 1889, fewer than 600 bison survived in the United States.

Several years later, the government outlawed bison hunting. Lands were set aside especially for the animals' protection. Slowly the bison population began to recover. Today there are more than 40,000 animals on preserves and on ranches.

The wisent (say VEE-zent), or European bison, is a taller relative of the North American animal. The chestnut-colored wisent lacks its American relative's long, shaggy hair and humped shoulders. The wisent once was also in danger of becoming extinct. Cities and towns had replaced many of the woodlands where it made its home. Now the wisent lives in zoos and on preserves.

Wisent nibbles on twigs on a preserve in Scotland. ▷
The wisent, a European bison, lacks the bulky mane and hump of its North American relative. The wisent lives in forests instead of on plains.

Wisent: 6½ ft (198 cm) tall at the shoulder

BISON

HEIGHT: 5-6½ ft (152-198 cm) at the shoulder

WEIGHT: 930-2,200 lb (422-998 kg)

HABITAT AND RANGE: prairies, plains, forests, and woodlands in North America and in Europe; most bison now live on preserves

FOOD: grasses, herbs, leaves, shrubs, and twigs

LIFE SPAN: 12 to 15 years on preserves

REPRODUCTION: 1 young after a pregnancy of about 9 months

ORDER: artiodactyls

Black buck
The black buck is a kind of antelope. Read about antelopes on page 52.

Boar
Boar is a name for a type of hog. Read about hogs on page 264.

Bobcat
(*say* BOB-cat)

Perched on a rock, a bobcat warms itself on a winter day in Colorado. People rarely see these animals because they hunt at night. During the day, bobcats usually stay hidden among rocks or thick brush.

THROUGHOUT FORESTS, mountains, swamps, and deserts, the bobcat roams and hunts. Wherever prey is plentiful in North America, this shy, strong animal can live and find food. Though only about twice as large as an ordinary house cat, a bobcat can kill an animal several times its own size. Perhaps that is why it is also known as a wildcat. Usually, the bobcat goes after smaller prey. It hunts rabbits, hares, mice, and squirrels. Crouching low to the ground, it slowly creeps toward its victim. Then, swiftly, it pounces. It may leap as far as 10 feet (305 cm) to catch an animal.

A bobcat's soft, tan fur is spotted with black. Fringes of long side-whiskers grow out from beneath its tufted ears. The bobcat's tail—with its black tip and white underside—is only about 6 inches (15 cm) long. In fact, the cat is named for its stubby tail, which seems to be cut off, or bobbed!

Bobcats usually live alone. They make their homes among rocks and bushes and in caves and hollow logs. The female chooses a hidden spot to use as a den. There she will give birth to her litter of one to six young.

After nursing her kittens for two months, the mother will begin bringing meat back to the den for them. A month or two later, she will begin to take them on nighttime hunts. As she leads her young around rocks and bushes, the mother raises her tail. Her kittens can see the white fur in the darkness.

After 9 to 12 months, the kittens are ready to leave the den and live alone. In the wild, they may survive about 10 to 12 years. Raised in a zoo, however, they might live to be 25 years old.

BOBCAT

LENGTH OF HEAD AND BODY: 26-41 in (66-104 cm); tail, 4-7 in (10-18 cm)

WEIGHT: 11-30 lb (5-14 kg)

HABITAT AND RANGE: forests, mountains, swamps, and deserts throughout most of North America

FOOD: hares, rabbits, rodents, and birds

LIFE SPAN: 10 to 12 years in the wild

REPRODUCTION: 1 to 6 young after a pregnancy of about 2 months

ORDER: carnivores

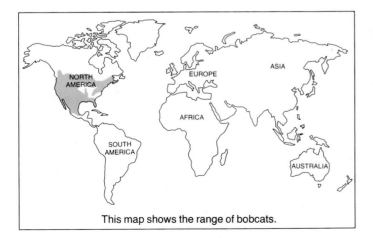

This map shows the range of bobcats.

Fluffy seven-day-old bobcat kittens (above) snuggle together in a den in a hollow log. The young have not yet opened their eyes. But in about two more days they will be able to see. Soon they will be climbing trees, like the two-month-old bobcat below. Full-grown bobcats are about twice as large as house cats. They can weigh as much as 30 pounds (14 kg).

▽ *Nine-month-old bobcat leaps as if pouncing on its prey. Actually it is only playing in the snow. Still as frisky as a young kitten, this bobcat soon will leave its mother. It will begin to live and to hunt on its own.*

Bongo
The bongo is a kind of antelope. Read about antelopes on page 52.

Brocket
The brocket is a kind of deer. Read about deer on page 170.

Buffalo

(say BUFF-uh-low)

SWIMMING IN A RIVER cools a buffalo on a hot afternoon. Sometimes this large, cowlike animal even wallows neck-deep in a mudhole. Because the buffalo has few sweat glands in its skin, it cannot cool off by sweating. It lowers its body temperature with a swim or a mud bath. These activities also keep pesky insects off the buffalo's tough hide.

A buffalo never strays far from rivers, creeks, or water holes. After its swim, it finds a shady spot in the underbrush and settles down to rest. During the night or in the early morning, buffaloes drink and graze. When feeding, an animal bites off blades of grass and swallows the partly chewed food. The food goes into its stomach, where it is formed into wads called cuds. After eating, the animal brings up a cud and chews it slowly. After chewing the cud thoroughly, the buffalo again swallows and finally digests it. Antelopes, camels, cows, deer, goats, and

Cape buffalo: 5 ft (152 cm) tall at the shoulder

△ *Huge handlebar horns curve from the head of a Cape buffalo. The contented animal ignores the tiny hitchhiker perched near its eye. The small bird—known as an oxpecker—eats ticks and insects from the buffalo's coat. If an oxpecker chirps and flies away, a buffalo becomes alert and watches for an approaching enemy.*

sheep also chew cuds. Read about these animals under their own headings.

Buffaloes have adapted, or become suited, to many environments. Buffaloes in Africa live on open grassy plains and in dense forests. In Asia, another kind of buffalo makes its home in the foothills of the Himalayas. Some buffaloes are even found on tropical islands in the Pacific Ocean.

No buffaloes live in North America, however. The animal that is sometimes called the American buffalo is actually the bison. Find out more about the bison on page 100.

Not all buffaloes look alike. One kind of buffalo

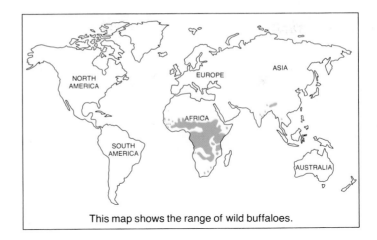

This map shows the range of wild buffaloes.

Cape buffaloes drink at a water hole in Africa. Buffaloes usually spend the night and early morning hours drinking and feeding. Later in the day, the animals wallow in the mud or seek shady places to rest.

may differ from another in height, weight, or color. But most of these animals have short, thick necks, broad heads, and long tails. Stout legs support their stocky bodies. Heavy, pointed horns curve outward from the heads of both males and females.

The Cape buffalo of Africa is well known among hunters. It has a reputation as the most dangerous big-game animal. This huge, coal-black buffalo measures more than 5 feet (152 cm) tall at the shoulder and weighs as much as 1,300 pounds (590 kg). When it charges, it may run 30 miles (48 km) an hour!

Lions sometimes prey on adult Cape buffaloes. Other hunters such as leopards and hyenas seek out wounded animals or the very young or old. When a Cape buffalo is under attack, however, another member of the herd may come to its rescue. Charging at top speed, a stronger male or female buffalo tries to drive off the enemy. The buffalo's thick, sharp horns—perhaps 4 feet (122 cm) from tip to tip—make dangerous weapons. By twisting and

Two-month-old Cape buffalo strolls with three ▷
oxpeckers on its back. The calf's woolly brown coat will grow darker and coarser with age. Its stubby horns will not reach their full size for five years.

▽ *Standoff! A male Cape buffalo defends another buffalo by confronting three lions. One buffalo often comes to the rescue of another. The animal's huge horns— 4 feet (122 cm) from tip to tip—can pierce the hide of an attacker. A herd of wildebeests grazes in the distance.*

charging, a buffalo can easily gore an attacker's body with its horns.

In the past, a herd of Cape buffaloes often numbered as many as 2,000 animals. In the late 1800s, however, a widespread cattle disease began to kill many African buffaloes. Until recently, herds were much smaller than those of a century ago. Today scientists have controlled the disease, and the size of the herds is increasing again.

Another kind of African buffalo, the forest buffalo, lives in smaller herds of about fifty animals. The forest buffalo makes its home in rain forests near the Equator. Smaller and lighter than the Cape buffalo, it measures nearly 4 feet (122 cm) tall at the

Resting in the midday heat, a Cape buffalo peacefully shares a mud bath with a group of hippopotamuses. Because the buffalo's skin has few sweat glands, the animal cannot cool off by sweating. Instead, it lowers its body temperature by wallowing in the mud.

shoulder and weighs about 660 pounds (299 kg). The forest buffalo has a thick, reddish brown coat. Its short horns sweep back, ending in sharp points.

Both the Cape buffalo and the forest buffalo wander free in their African homelands. In Asia, however, few buffaloes roam wild. People began to domesticate, or tame, these animals about 5,000 years ago. Domestic buffaloes pull plows and carts. Often they carry heavy loads on their backs. Some also are raised for their milk or meat.

The best known buffalo in Asia is the stocky

BUFFALO

HEIGHT: 3-5 ft (91-152 cm) at the shoulder

WEIGHT: 450-1,500 lb (204-680 kg)

HABITAT AND RANGE: forests, woodlands, grasslands, swamps, and mountains in Asia and in Africa south of the Sahara; domestic buffaloes live in Asia, Egypt, Europe, and South America

FOOD: grasses and leaves of shrubs

LIFE SPAN: 16 to 20 years in the wild

REPRODUCTION: 1 young after a pregnancy of 9 to 11 months

ORDER: artiodactyls

Buffalo

water buffalo. This barrel-shaped animal measures about 5 feet (152 cm) tall at the shoulders. Its enormous, curved horns—the largest of any buffalo—may reach 5 feet (152 cm) from tip to tip. A water buffalo's nearly hairless coat is usually dull gray or black. Some breeds that live on islands in the Pacific Ocean have white coats.

True to its name, the water buffalo often lies in a river or a creek from morning until evening. Only its head and horns show above the water's surface. Often a water buffalo will roll and wallow in mud at the water's edge. It covers its skin with the mud, protecting itself from insects.

Thai boy balances on a female water buffalo's ridged ▷
back while her year-old calf wades nearby. Small children often tend these obedient animals.

▽ *Herders drive water buffaloes through high water in Brazil. These hardy animals, originally from Asia, now live in many parts of the world.*

△ *Two-month-old calf huddles close to its mother among a herd of water buffaloes in India. A few small herds like this one roam wild in parts of Asia.*

Neck-deep in a river, a pair of Asian water ▷ *buffaloes takes a cooling swim after a morning spent plowing rice fields. Asians have used buffaloes as work animals for centuries. Buffaloes also work in fields in parts of Europe.*

Water buffalo: 5 ft (152 cm) tall at the shoulder

Another kind of Asian buffalo, the small tamarau (say tam-uh-RAU), lives in swampy areas of mountain forests on one of the Philippine Islands. It measures about 3½ feet (107 cm) tall at the shoulder. The tamarau's coat is dark gray with white marks on its head, neck, and legs. The animal's horns are ridged and slightly curved.

The smallest kind of Asian buffalo, the anoa (say uh-NO-uh), makes its home in Indonesia. This brown buffalo grows only about 3 feet (91 cm) tall. Its small, straight horns point backward from the top of its head. The anoa lives in wooded mountains. The rarely seen animal stays in thick underbrush, and little is known about its habits.

A female buffalo—African or Asian—may bear a single calf each year. A pregnancy varies in length from nine to eleven months, depending on the kind of buffalo. The newborn stands within thirty minutes of its birth. For two years, it stays close to its mother in the herd. But it may join other calves to play. The frisky calves chase one another, tease their mothers, or run after their own tails. Although the calves grow rapidly during their first year, they do not reach adult size until about the age of five.

Burro
The burro is a domestic ass. Read about asses on page 66.

Bush baby

THE LOUD, SHRILL CRIES of the tiny bush baby helped earn the animal its name. When it calls out, this member of the primate order sounds surprisingly like a human baby. Other primates include monkeys, apes, and humans.

Bush babies live in Africa, in forests and in open shrubby areas called the bush. The largest kinds are about the size of cats. Others are only as big as chipmunks. All have long tails. Pads on their fingers and toes help them cling to trees.

Smaller bush babies are expert jumpers. They seem to fly among the branches, but actually they are leaping. Using their muscular legs, they spring from tree to tree. On the ground, they can hop like tiny kangaroos.

Bush babies spend the day resting. Sometimes they sleep in nests made of leaves and twigs. Or they may curl up in hollow trees or in tree forks. At night, the furry animals move about searching for food. Bush babies eat tree gum, insects, lizards, mice, and small birds. Larger kinds of bush babies eat fruit to fill out their diet.

With its big eyes, a bush baby can spot prey in the dark. Sensitive hearing alerts it to the approach of another animal. The bush baby moves its ears in the direction of a noise. Even the slightest sound will not escape its keen hearing. Before sleeping, the animal folds up its ears like fans.

Females usually give birth to one or two young after a pregnancy of about four months. The offspring are raised in nests made of leaves. Frequently a mother will move her young to another nest. She carries them in her mouth, or they cling tightly to her fur. Sometimes she will leave, or "park," her infants in different spots for a short time. She may even park them for most of the night while she searches for food. But she brings them back to the nest before dawn.

Bush babies are popular pets in some parts of the world. Tame ones may lick their owners' faces and ride about in their pockets.

Another name for the bush baby is galago (say guh-LAY-go). Find out about its close relatives the loris on page 348 and the potto on page 459.

Seeking security, young Senegal bush babies huddle in a tree. These African animals often sleep clustered together.

Senegal bush baby: 6 in (15 cm) long; tail, 9 in (23 cm)

Senegal bush baby peers out from a tree fork. Big eyes help this animal see well in the dark. Keen hearing alerts it to the approach of prey. The bush baby can move its ears in the direction of a sound. It also can fold them up like fans. The bush baby sleeps with its ears curled shut.

Thick-tailed bush baby: 13 in (33 cm) long; tail, 15 in (38 cm)

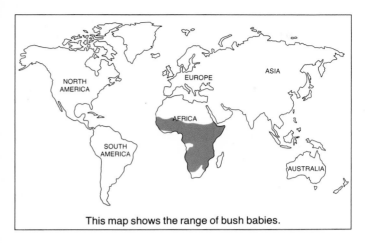

This map shows the range of bush babies.

BUSH BABY

LENGTH OF HEAD AND BODY: 5-14 in (13-36 cm); tail, 7-16 in (18-41 cm)

WEIGHT: 11 oz-4 lb (312 g-2 kg)

HABITAT AND RANGE: forests and brushy regions of Africa

FOOD: tree gum, fruit, insects, and other small animals

LIFE SPAN: 10 to 14 years in captivity

REPRODUCTION: usually 1 or 2 young after a pregnancy of about 4 months

ORDER: primates

Thick-tailed bush baby crouches on a branch. Pads on the tips of its fingers and toes help it grip the branch. When this furry animal senses danger, it scampers away.

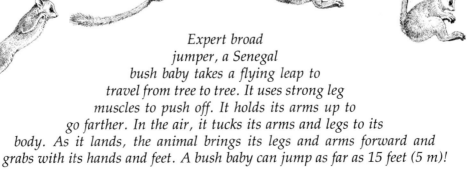

Expert broad jumper, a Senegal bush baby takes a flying leap to travel from tree to tree. It uses strong leg muscles to push off. It holds its arms up to go farther. In the air, it tucks its arms and legs to its body. As it lands, the animal brings its legs and arms forward and grabs with its hands and feet. A bush baby can jump as far as 15 feet (5 m)!

C

Cacomistle
The cacomistle is a close relative of the ringtail. Read about ringtails on page 494.

Camel
(*say* KAM-ul)

UNDER THE DESERT SUN, merchants lead a caravan of camels across hot sand. The camels will drink no water until they reach the next oasis—perhaps three or four days later. Yet these hardy animals keep up a steady pace. They may travel 100 miles (161 km) without water.

Camels do not easily sweat. Therefore they lose the moisture in their bodies slowly. They get the moisture they need by drinking water and by eating desert plants. In winter, plants provide enough moisture for camels to go without drinking for several weeks!

Even when water is available at wells and at water holes, camels drink only if necessary. They take in just enough to replace the water used since their last drink. Sometimes, however, they have used up quite a lot. A thirsty camel can gulp down as much as 30 gallons (114 L) of water in just ten minutes! That would be like drinking 480 cups of water in the same length of time.

Arabian, or one-humped, camels are found mainly in the hot deserts of North Africa and Asia.

Bearing bags of salt to market, a caravan of Arabian camels winds across the Sahara in Africa. These hardy pack animals can carry heavy loads about 25 miles (40 km) a day. Some desert peoples measure wealth by the number of camels a person owns.

They live in regions where temperatures rise above 120°F (49°C). Their short coats help to block out the heat of the sun. With broad, thickly padded feet, they walk easily on shifting sands. Arabian camels stand more than 7 feet (213 cm) tall at the hump and weigh as much as 1,600 pounds (726 kg). People sometimes call Arabian camels dromedaries (say DRAH-muh-dare-eez).

Bactrian (say BACK-tree-un) camels have two humps. They also have very tough feet for crossing the rocky deserts of Asia. Temperatures there range from a low of minus 20°F (minus 29°C) in winter to more than 100°F (38°C) in summer. The thick,

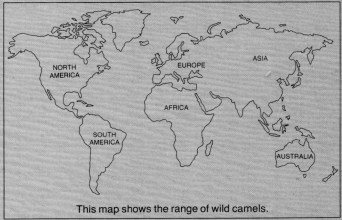

This map shows the range of wild camels.

Protected from the blistering heat by short hair and padded hooves, Arabian camels cross a ridge of sand in a desert in Africa. The animals range in color from brown to "camel's-hair" tan to white.

shaggy coats of these camels protect them from the cold. When temperatures soar, they shed their heavy winter coats. Stockier and slightly shorter than their Arabian relatives, Bactrian camels measure about 7 feet (213 cm) tall at the hump and weigh about 1,800 pounds (816 kg).

Well suited to life in harsh deserts, both Arabian and Bactrian camels have bushy eyebrows and double rows of eyelashes. Their ears are lined with hairs. Special muscles allow them to close their nostrils and lips tightly for long periods. These features help protect them from blowing sand or snow.

Camels' humps help the animals survive in the desert. Many people think that camels store water in their humps, but the humps are really masses of fat. This fat nourishes the animals when food is scarce. With this energy supply on their backs, the animals can go several days without eating. Camels store about 80 pounds (36 kg) of fat in their humps. As camels use this fat, their humps shrink. If camels do not eat, their humps get flabby. The humps become firm again after the camels eat and drink.

Camels nibble at whatever plants they can find. Like cows, they do not chew their food completely

CAMEL

HEIGHT: 6-7½ ft (183-229 cm) at the hump

WEIGHT: 1,000-1,800 lb (454-816 kg)

HABITAT AND RANGE: the Gobi in Asia; domesticated camels live in Africa, Asia, the Middle East, and Australia

FOOD: grasses, juicy plants, leaves, branches, grains, and dates

LIFE SPAN: 25 to 30 years in captivity

REPRODUCTION: 1 young after a pregnancy of about 12 or 13 months

ORDER: artiodactyls

▽ *Bactrian camel moves quickly across a plain. One of its humps has flopped sideways, because the animal has used the fat stored inside it for energy.*

Bactrian camel: 7 ft (213 cm) tall at the hump

before swallowing it. After eating, they bring up the partly digested food, called a cud, and chew it thoroughly. Then they swallow the cud and digest it.

Camels' mouths are so tough that even the sharp thorns of desert plants do not hurt them. If they cannot find food, camels nibble on ropes, on sandals, and sometimes on their owners' tents!

A female camel may give birth to a calf every other year. The newborn can stand shortly after birth and can walk within a few hours. It stays with its mother until it is almost two years old. It is not fully grown, however, until the age of five.

People first domesticated, or tamed, camels at least 3,500 years ago. Today almost all of them are domestic. Of the millions of camels in the world, probably fewer than a thousand roam wild.

For desert peoples, camels are strong and dependable work animals. Camels carry goods to market or carry riders. The animals also provide their owners with food, clothing, and fuel. Camel milk is thick and rich, and people sometimes eat camel meat. The fat inside their humps can be melted down and used for cooking.

When the animals shed their coats, their owners gather up the woolly hair. They weave it into clothing, blankets, and tents. From the tough hide, they make shoes and saddles. They burn dry camel waste as fuel for cooking and heating.

Capybara

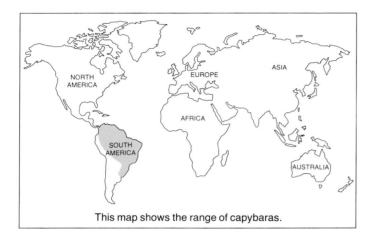

This map shows the range of capybaras.

WORLD'S LARGEST RODENT, the capybara is at home on land or in the water. If an enemy like a jaguar or a person chases it, the animal can run or leap away. It is an able swimmer. In the face of danger, capybaras will plunge into water. Only their eyes, ears, and nostrils show as they paddle away. Their webbed toes act a little like swim fins. The animals can also stay underwater for several minutes.

Capybaras look somewhat like giant, long-legged guinea pigs. They grow more than 4 feet (122 cm) long and weigh as much as 110 pounds (50 kg). Their coarse hair is so thin that their skin can dry out in the hot sun of Central and South America where

In a muddy river in Brazil, a capybara provides a hairy perch for a cattle tyrant bird. Capybaras often wallow in water. This keeps their skin from drying out in the hot sun.

they live. So capybaras often spend time wallowing in the mud. They also rest in the shade of swampy woods. In the morning and evening, they stand in water up to their stomachs and feed on plants. The front teeth of capybaras, like those of all rodents, keep growing. Capybaras must chew and gnaw all their lives to wear down their teeth.

Capybaras live in groups of about twenty adults. A female capybara has a litter of one to six young each year. These offspring—born with hair and able to see—weigh about 2 pounds (1 kg). The young follow their mother for a year.

Capybaras do not dig burrows. Instead they hollow shallow beds in the ground. When they live near gardens and ranches, the animals sometimes eat melons, squash, corn, and other crops. For this reason—and for their meat—people in South America often hunt capybaras.

CAPYBARA

LENGTH OF HEAD AND BODY: 40-52 in (102-132 cm); height at the shoulder, 20 in (51 cm)

WEIGHT: 60-110 lb (27-50 kg)

HABITAT AND RANGE: marshes, swamps, and wooded areas near rivers and lakes in Central and South America

FOOD: aquatic plants, grains, and fruit

LIFE SPAN: 8 to 10 years in the wild

REPRODUCTION: 1 to 6 offspring after a pregnancy of about 4 months

ORDER: rodents

◁ *Dripping wet, a young capybara climbs onto a riverbank in Brazil. Capybaras weigh about 2 pounds (1 kg) at birth. The newborn have hair and can see.*

▽ *Capybaras gather near water during the dry season. At other times the animals live in smaller groups. The animals communicate with grunts and whistles.*

Caracal

The caracal is a kind of cat. Read about caracals and other cats on page 126.

Caribou

(*say* CARE-uh-boo)

AS SNOW BEGINS TO MELT in early May, caribou of Alaska and Canada start to travel, or migrate, to their summer feeding grounds. Female caribou—called cows—move about 400 miles (644 km) north along trails worn smooth over years of migrating.

Day and night, the herd flows along. Thousands of caribou feed, rest, and move on again. Male caribou—called bulls—follow a few weeks later, along with yearlings, calves born the year before.

Broad hooves help caribou cross the harsh land. The hooves act like snowshoes in deep snow and like paddles in water. Sharp edges help caribou get good footing on rocky hillsides and on slick ice. With the scoop-shaped undersides of their hooves, caribou can dig through the snow to find food. Each animal clears many feeding holes every day.

As they travel, caribou shed hair from their thick winter coats. Heaps of matted grayish brown hair mark places where many animals have passed.

By June, the cows and their calves, born along the way, have arrived at their summer pastures. A

Crowned with sharp-tipped antlers, a male caribou— ▷ called a bull—rests in Alaska's September sunshine.

▽ With their month-old calves nearby, female caribou— called cows—graze in a summer feeding ground.

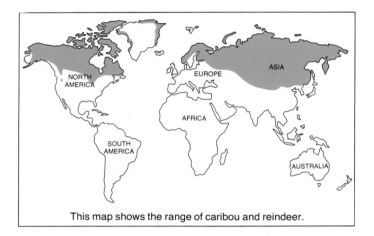

This map shows the range of caribou and reindeer.

△ *Shreds of velvet, a covering of soft skin, hang from the antlers of a caribou bull in early fall. Though the pieces look bloody, shedding the velvet does not hurt the animal.*

On the slopes of Mount McKinley, caribou bulls roam ▷ their summer grounds. They feed in one valley after another, wandering all the time.

cow usually has one offspring a year. Wolves, wolverines, lynxes, and bears often prey on the newborn animals. But the calves develop quickly. A young caribou can stand on its feet a few minutes after birth. The next day it can follow its mother as she looks for food.

New calves nurse often at first. But after six months, they eat only plants, as adults do. In cool summer pastures, they feed on grasses and on the leaves of shrubs and other plants. An adult eats about 12 pounds (5 kg) of food each day.

Summer brings huge swarms of insects. Mosquitoes and flies bite the animals' bodies. Some flies lay eggs under their skins. The frantic caribou splash in ponds or climb to breezy hilltops to ease the itching and to escape from the insects.

By July, the bulls have grown new sets of antlers covered with a soft skin called velvet. Bulls grow and shed their antlers every year. The new antlers begin as bumps. These grow and branch out.

In early fall, the velvet begins to fall off. To help remove it, a bull rubs his antlers against bushes. This exposes the sharp points of bone. During the fall mating season, bulls use their antlers to jab and to wrestle with each other as they fight for mates. Just before winter, their antlers fall off.

Caribou cows have smaller and thinner antlers. They begin to grow in late summer, and they drop off the next June. Deer, elk, and other close relatives of caribou also grow and shed antlers—but only the males. Read more about antlers on page 170.

With the first snowfall, the entire herd—bulls, cows, and calves—begins to head south to spend the winter in sheltered woodlands. Then, in early May, caribou begin their migration north again.

For centuries, the Eskimo have depended on caribou meat for food. They have used hide for clothing, tents, and kayaks. They have made needles and fishhooks from caribou bones and antlers.

Even today, in northern Europe and Asia, people herd another kind of caribou, called the reindeer. They follow their herds on skis, by boat, and in snowmobiles. They eat reindeer meat. And they make cheese and butter from reindeer milk.

Caribou move north across a snowfield in Alaska. Sharp ▷ hooves help them cross the terrain, still icy in late spring.

CARIBOU

HEIGHT: 4-5 ft (122-152 cm) at the shoulder

WEIGHT: 240-700 lb (109-318 kg)

HABITAT AND RANGE: tundra, northern forests, and mountain uplands from western Alaska through Canada to western Greenland, northern Europe, and northern Asia

FOOD: mostly grasses and tiny plants, and some leaves

LIFE SPAN: about 15 years in the wild

REPRODUCTION: 1 young after a pregnancy of about 8 months

ORDER: artiodactyls

△ With bare antlers, two young bulls spar. They practice for the time when they will fight for a mate. Each animal tilts his head sideways and jabs with the sharp points. After the mating season, the antlers will fall off.

◁ Caribou cross the chilly Kobuk River on their yearly migration. The hollow hairs of their outer coats trap air, helping the animals float. Bulls wear heavy antlers. Cows, with delicate sets, swim alongside and behind. Caribou migrate as many as 800 miles (1,287 km) each year.

125

Cat

CATS ARE FULL OF CONTRASTS. Their soft, padded paws hide sharp, hooked claws. At rest, cats stretch out lazily. On the move, they glide silently through the grass. Their glossy fur covers powerful muscles. A cat can freeze, still as a statue, then explode in a savage attack. There are more than 35 different species, or kinds, of cats. All the different breeds of house cats make up only one of these species. The rest are wild cats. Under their own headings, you can read more about some of the better-known wild cats: bobcat, cheetah, jaguar, leopard, lion, lynx, and tiger.

Cats vary in size — from a dainty pet calico cat to a huge Siberian tiger. They vary in color — from a pure white Persian cat to a velvety black leopard.

Regal and watchful, two young mountain lions rest in a cave in Utah. The mother of the cubs relaxes in the shadows. Mountain lions often sleep during the day and hunt at night.

This map shows the range of wild cats.

Mountain lion: 53 in (135 cm) long; tail, 27 in (69 cm)

But members of the cat family are not hard to recognize. Most cats have rounded heads with rather flat faces. Their sleek, streamlined bodies move gracefully on muscular legs.

Cats live in almost every kind of environment throughout the world. The snow leopard pads along the cold, rocky slopes of the Himalayas, the highest mountains on earth. The ocelot (say AH-suh-lot) prowls through steamy South American jungles. The caracal (say CAR-uh-kal) roams dry, desert country in Africa and Asia. And the rare Iriomote (say ear-ee-uh-MO-tay) cat lives deep in a forest on only one small island in Japan.

Cats and people have lived together for centuries. At first, wild cats probably hunted the mice and rats that lived where food was stored. People began to like these mousers and treated them well. Gradually, cats moved into people's homes.

Cats probably were first tamed in Egypt about 4,000 years ago. The ancient Egyptians worshiped a cat goddess. When a family cat died, people would cut their hair to show how sad they were. Thousands of dead cats were made into mummies. They were even supplied with mouse mummies for food in the next world!

Most cats are good at running, jumping, climbing, and even swimming. Some are star performers. The rippling muscles in a mountain lion's hind legs can send it soaring through the air to pounce on its prey or to sail over an obstacle in its path. As a serval

◁ *Roaming the brushy hills of Idaho, a young adult mountain lion looks over its home range. These cats often live in remote, dry areas with deep canyons and steep cliffs. There they prey mainly on deer, elk, and hares.*

▽ *Male mountain lion (below) carefully licks his claws to clean them and to remove bits of meat, bone, and hair. When not in use, a mountain lion's claws retract into protective coverings in its paws. At bottom, another mountain lion prowls through the snow on padded feet. A long, graceful tail helps it keep its balance while stalking, running, leaping, and climbing.*

(say SIR-vul) hunts, it moves through tall grass in fantastic flying leaps. A fishing cat uses the weblike skin between its toes to help it swim. Sometimes it even dives underwater.

A cat walks on pads that are under the toes and the sole of each foot. It seems to walk on tiptoe. When its claws are not in use, they retract into protective coverings. The long, curved claws always stay needle-sharp.

All cats, tame or wild, are effective hunters. In the wild, cats prey on almost anything that moves, from insects to buffaloes. A caracal can leap up and snatch a bird right out of the air. With a swift move of its paw, a fishing cat can scoop fish, crabs, and frogs out of a stream. Smaller cats feed mostly on mice, rats, and birds. Big cats eat antelopes and other larger animals, as well as small prey.

Arching leaves surround a frosty-looking Pallas's cat. ▷
These animals have the longest and thickest fur of all wild cats. Their coats of white-tipped hairs protect them from the harsh weather of their homeland in central Asia.

▽ *With a swift and sure leap, a serval pounces toward its prey. The tall grass of eastern Africa hides many small animals. So servals hunt mostly by sound rather than by sight. They usually eat snakes, birds, mice, and rats.*

No matter what they hunt, all cats get their food in a similar way. Usually they hunt alone. Sometimes they hide and wait for an animal to pass by. More often they silently creep up on their prey.

When they stalk another animal, cats move swiftly at first. Then they suddenly stop and watch intently. They stay flat against the ground and under cover. They creep again and stop again. As they get closer, they move even more slowly. Some cats

Pallas's cat: 22 in (56 cm) long; tail, 10 in (25 cm)

Serval: 32 in (81 cm) long; tail, 16 in (41 cm)

Ocelot: 40 in (102 cm) long; tail, 12 in (30 cm)

△ *Surefooted ocelot tiptoes along a dead branch in Venezuela. Ocelots usually hunt on the ground. But sometimes they stalk birds or monkeys in the trees.*

Pampas cat: 25 in (64 cm) long; tail, 12 in (30 cm)

△ *Spots and stripes mark the fur of a pampas cat. These markings help it blend into its many habitats: grasslands, mountains, and forests of South America.*

◁ *Stalking jungle cat barely disturbs the morning calm. Jungle cats live throughout the Middle East and Asia. They feed on hares, rats, mice, and lizards. Centuries ago, Egyptians trained jungle cats to hunt birds for them.*

Jungle cat: 26 in (66 cm) long; tail, 11 in (28 cm)

131

Quick and skillful, a fishing cat wades into the water to catch its dinner. On a sandbank in a river in India, this rarely seen cat arches its back and pounces on a fish. It pins its prey to the streambed with its front paws.

crouch so low that their shoulder blades and hip bones stick up above their backs. When they are very close to their prey, cats spring. They seize prey with sharp teeth and strong claws.

Cats kill their prey with a well-aimed bite of their powerful jaws. An ancient relative of modern cats was the saber-toothed cat. Its long fangs could pierce the toughest hides. Today cats still have pointed fangs. They use these teeth to bite prey. Their sharp-edged back teeth cut through meat like scissors. Cats have no flat-topped teeth for chewing food, so they cannot grind their meat. Cats swallow their food without chewing.

A cat's tongue is rough, like sandpaper. It can

CAT

LENGTH OF HEAD AND BODY: 13 in-6 ft (33 cm-183 cm); tail, 6 in-3 ft (15 cm-91 cm)

WEIGHT: 4-500 lb (2-227 kg)

HABITAT AND RANGE: every kind of habitat worldwide, except in Antarctica, Australia, Madagascar, the West Indies, and some oceanic islands; domestic cats are found almost everywhere

FOOD: animals and sometimes plants

LIFE SPAN: 12 to more than 20 years in captivity, depending on species

REPRODUCTION: 1 to 8 young after a pregnancy of about 2 to 4 months, depending on species

ORDER: carnivores

lick a bone clean of the last shred of meat. Cats also use their tongues to clean themselves. By bending and stretching, a cat can reach almost every hair on its body with its tongue.

Of all the senses, cats depend most on their sight and hearing. A cat's ears turn to catch the slightest sound. The serval has enormous oval ears. It is hard for this cat to see prey in the tall grass of the African plains where it lives. But the serval's huge ears pick up every chirp, squeak, and rustle nearby.

The Pallas's (say PAL-us-ez) cat lives in the cold, harsh areas of central Asia. Its ears are very short. They are low on its head and wide apart. The eyes of this cat are very near the top of its head. This arrangement may help the cat hunt where there are few bushes to hide behind. It can peer over rocks and still not show much of itself.

Cats do most of their hunting in the dark. Their eyesight at night is good. Like other nighttime hunters, they can see when it is dark, because inside a cat's eye there is a special surface that reflects light. This is what makes a cat's eyes shine in the dark. Many cats have pupils that become tall, narrow slits in bright light. In the dark, these pupils open wide to let in light.

Cats also find their way in the dark with their whiskers. These long, stiff hairs on their faces have

Then it rears up on its two back legs, gripping the struggling fish in the air. After killing it, the cat again places the fish underwater. Many cats do not like the water, but the fishing cat usually lives near streams or rivers.

no real feeling. But if the tips of the whiskers brush something, a cat feels the movements at the roots of its whiskers.

Cats use their sense of smell to communicate. They mark trees or lookout points by scratching them with their claws and by leaving their waste. These places are called scent posts. By their odor, they show where a cat's home range is.

A long, graceful tail helps a cat keep its balance as it leaps and climbs. The movement of the tail can indicate a cat's mood—excitement, anger, fright, relaxation. On some cats, markings make the tail easier for others to see. Tails of jungle cats and pampas (say PAM-pus) cats are ringed and tipped with black.

Most cats have coats of dark spots on lighter fur. A snow leopard's spots look a little like large roses. An ocelot's spots resemble links in a chain. The background color of most wild cats is light brown or golden yellow. But the snow leopard has gray fur. And the serval's fur may be reddish.

A cat's color and markings help it hide while creeping up on prey. The snow leopard blends into its rocky background. The brown Iriomote cat seems to disappear among forest shadows. The spots and stripes of the pampas cat help it hide in tall grass. And the tan caracal matches its sandy home.

Cats that live in warm climates generally have

The cat moves the fish to a safer hold in its mouth. It carries its prize this way as it crosses a sandbank and then goes through deeper water to shore.

shorter fur. In cool climates, fur is longer. The Pallas's cat lives in high, cold regions. Some people think that the extremely long, thick fur on its underside protects it when it lies on the frozen ground. But short, thin fur on its back may let this cat soak up heat from the sun.

House cats have been bred to suit different people. So their fur may be long, short, or even curly!

◁ Female snow leopard, prized for its spotted coat, peers over a ridge in the rugged mountains of Pakistan. The cat roams widely looking for such prey as wild sheep. It comes down from the high, treeless slopes only in winter.

Snow leopard: 41 in (104 cm) long; tail, 35 in (89 cm)

It may be striped, spotted, marbled, or plain. And besides the more common shades of tan, brown, and black, their fur may be silver, blue, or ginger.

Different kinds of cats mew, purr, snarl, growl, grunt, hiss, yowl, scream, and roar. But no cat makes all these noises. Only some of the big cats can roar. And the bigger the cat, the louder the noise. Smaller cats can purr for a long time. When a big cat purrs, it has to stop to catch its breath. Cats' piercing yowls and screams often can be heard for long distances. These sounds, along with scents, help cats mark their ranges and find partners at mating time.

In cool climates, wild cats mate in winter or in early spring. The young are then born two to four months later. In warm climates, the young may be born at any time of year.

The female chooses a hidden spot for a den. A mountain lion may find a cave. An ocelot looks for a hollow tree. A serval may take over a porcupine's burrow. A jungle cat may hide among dry reeds.

Some cats have only one or two young at a time. Others have three or four. Tame cats sometimes have eight or more helpless little balls of fur. Young cats are called cubs or kittens. They are born with their eyes closed, and they are barely able to move.

The mother feeds her kittens, cleans them, and protects them. If danger threatens, she may move her litter to a new nest. A female cat carries her kittens, one at a time, in her mouth. She holds them gently by the neck and shoulders. Females almost always care for their young alone. Most mothers will not let males get near the young.

Cubs soon start exploring the world around their den. They play at hunting right away. On wobbly legs, one cub follows the trail of a fluttering leaf. Another rears up to catch its mother's twitching tail. From a rock, one cub leaps onto the back of another, and the two go tumbling in the dust. Cats often play games of stalk-and-pounce. As they grow older, they join their mother on hunting trips, learning the skills they will need later. Then most cats leave to find hunting grounds of their own.

▽ Long ears perked to catch sounds, a caracal watches its desert surroundings. The cat can move its ears to show confidence, anger, or alertness. Caracals roam from India through the Middle East and into Africa.

Caracal: 28 in (71 cm) long; tail, 9 in (23 cm)

Iriomote cat: 23 in (58 cm) long; tail, 8 in (20 cm)

△ Iriomote cat in Japan creeps through the nighttime shadows. It cocks its ears to catch a rustle in the forest. This rare cat hunts birds, snakes, and lizards. Until the mid-1960s, no one knew that such a cat existed!

Seal Point Siamese: 18 in (46 cm) long; tail, 12 in (30 cm)

Bi-color Persian: 18 in (46 cm) long; tail, 12 in (30 cm)

136   Red Tabby Shorthair: 18 in (46 cm) long; tail, 12 in (30 cm)

HOUSE CATS *have lived with people for centuries as popular and useful pets. More than thirty different breeds of tame cats exist. They vary a great deal—in color, markings, and length of fur. But all house cats belong to a single species, or kind, of cat.*

◁ *One of the most popular breeds of house cats, a Seal Point Siamese nestles among forget-me-nots.*

Scottish Fold: 18 in (46 cm) long; tail, 12 in (30 cm)

Cream Persian: 18 in (46 cm) long; tail, 12 in (30 cm)

△ *Fiery orange eyes blazing, a champion Cream Persian cat scowls. This breed of popular show cat has a flat muzzle and long, soft hair.*

Colorful patchwork of black, orange, and white covers ▷ *the coat of this Calico Rex. Its short hair grows in rippling waves from the top of its head to the tip of its tail.*

◁ *Black-and-white Bi-color Persian (far left) displays the long glossy fur and large round eyes typical of its breed. A Red Tabby Shorthair (left) stalks along the branches of a plum tree. Like their wild relatives, house cats hunt well. Kittens even pounce on insects and balls of dust.*

◁ *Pure white Scottish Fold cat shows off its eyes of different colors—one blue and the other deep coppery orange. Many white cats have these "odd eyes." A pink nose also often appears in light-colored cats. Cat shows provide good places to see how widely cats vary.*

▽ *Proud Abyssinian cat gazes at the world with large, almond-shaped eyes. This cat's fur looks grayish brown. But actually each hair in its coat has bands of black, brown, and white. People call this coloring "ticking."*

Abyssinian: 18 in (46 cm) long; tail, 12 in (30 cm)

Calico Rex: 18 in (46 cm) long; tail, 12 in (30 cm)

Cavy

The cavy is a close relative of the guinea pig. Read about both animals on page 247.

Chamois

(say SHAM-ee)

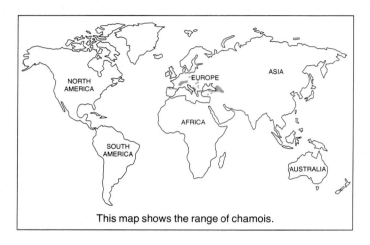

This map shows the range of chamois.

CHAMOIS

HEIGHT: 28-33 in (71-84 cm) at the shoulder

WEIGHT: 53-110 lb (24-50 kg)

HABITAT AND RANGE: mountains of Europe, southwestern Asia, and New Zealand

FOOD: grasses, herbs, flowers, evergreen shoots, woody plants, and dry leaves

LIFE SPAN: as long as 24 years in the wild

REPRODUCTION: 1 or 2 young after a pregnancy of about 6 months

ORDER: artiodactyls

LEAPING FROM ONE NARROW LEDGE to another, a chamois quickly climbs a steep mountain slope. It jumps gracefully across a 20-foot (6-m) ravine and lands lightly on the other side. The chamois is at home in the rugged mountains of Europe and of Asia. People have taken the animal to New Zealand, and it now lives in the wild there. It is so surefooted that it can balance on a small knob of rock.

The scientific, or Latin, name for chamois means "rock goat." In fact, the chamois looks like a goat, and it is closely related to the goat family. A male chamois measures about 30 inches (76 cm) tall at the shoulder and weighs about 90 pounds (41 kg). Both males and females have black, ringed horns. The chamois usually grows a new set of rings on each of its horns in the summer. If it has five sets of rings on each horn, it is probably five years old.

In summer, chamois feed on grasses, wild herbs, and flowers high in the mountains where no trees grow. When winter comes, chamois may come down to the forests below. There they feed on shoots, woody plants, and dry leaves.

Female chamois, called does, roam together with their young in herds of ten to thirty animals. Mature males, called bucks, usually wander alone. During the late spring and summer, the bucks pick out the best feeding areas for themselves. They eat a great deal and grow strong and fit. Groups of younger males and groups of females feed elsewhere.

At mating time in the late fall, the bucks join the females. Glands on each buck's head, near the base of his horns, produce a waxy substance. The buck rubs this substance on trees, bushes, and tall grasses to announce his presence. Chamois bucks fight over the females. They chase each other wildly along mountain slopes. As they race, they may try to jab each other with their hooked horns. The battle lasts until one buck runs off. The winner then looks for mates among the females.

About six months after mating, a female chamois gives birth, usually to one kid. Sometimes she may bear twins. The kids can walk almost immediately. Young chamois play much of the time, running and jumping and sliding down snow-covered hills. A kid follows its mother everywhere and stays close by her side.

As a herd of chamois grazes on a hillside, most members stay alert. If an enemy comes near, several chamois warn the rest of the herd with high-pitched whistles and the stamping of hooves. The chamois usually disappear among the rocks.

Eagles, lynxes, and wolves prey on chamois. People hunt them as game. Chamois skins are made into a very fine, soft leather. In most areas, laws now protect the chamois.

Warm in its winter coat, a male chamois climbs a ▷ snowy slope in Italy. A long, thick layer of hair protects the chamois from cold mountain winters. In summer, it sheds this coat for a shorter, lighter-colored one.

Cheetah

(*say* CHEE-tuh)

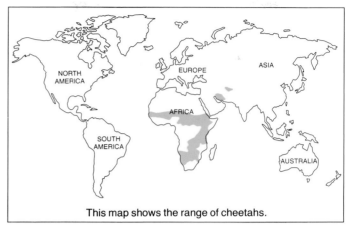

This map shows the range of cheetahs.

FASTEST MAMMAL ON LAND, the cheetah can reach speeds of 60 or perhaps even 70 miles (97-113 km) an hour over short distances. It usually chases its prey at only about half that speed, however. The cheetah can make sudden, sharp turns. With its long, muscular legs and flexible spine, it is the champion sprinter among land mammals.

Standing watchfully on a fallen acacia tree in Africa, a cheetah searches for prey. Because of its keen eyesight, this cat can see long distances across the grasslands. A short mane on the neck of the cheetah below marks the animal as young. The mane will disappear as it grows older.

△ *Sprawled atop a termite mound, a mother cheetah watches over her playful cubs. They stalk, run, and pounce. These games prepare them for the time when they must hunt alone for the animals they eat.*

Like other wild cats, the cheetah hunts to survive. Its excellent eyesight helps it find such prey as hares and antelopes during daylight hours. Though it rarely climbs trees, it sometimes perches on high places—a fallen tree, a hilltop, or a termite mound—and watches for prey.

When it sights prey, the cheetah usually begins to stalk. It creeps as close as possible before the attack. It may lift its head high to keep the prey in sight. But it keeps its body hidden. The cheetah is hard to see because its spotted coat blends with the tall, dry grass of the plains.

Suddenly, the cheetah makes a lightning dash. With a paw, it knocks its prey to the ground and then bites its throat. Usually, the cheetah drags the kill off to a shady spot. There it eats its meal.

Because it tires quickly, the cheetah does not always catch its victim. The cheetah can run at top

▽ *Young cubs huddle close to their mother for protection from enemies and for shade from the tropical sun. She will nurse them for several months, until they can eat meat.*

Lapping at a water hole, a cheetah drinks after a kill. ▷
Cheetahs drink only once every three or four days.

speed for only about 300 yards (274 m)—the length of three football fields. An antelope may be able to zigzag and get away from a cheetah.

A cheetah has claws on all four feet. Unlike other cats, the cheetah has no protective coverings for its claws. Only one claw on each front foot is sharp. This dewclaw is higher than the other claws and never touches the ground. Cheetahs use their dewclaws to knock down prey.

143

△ Fastest mammals on land, cheetahs make a lightning dash after a gazelle. They can briefly reach speeds of 60-70 miles (97-113 km) an hour. As they run, they knock their prey to the ground.

◁ Successful hunter, a cheetah drags a gazelle to the shade. It tries to keep its prey away from hyenas and vultures.

Cubs tear into their meal while ▷ their mother rests from the chase. Young cheetahs often watch their mother hunt. After about a year of watching and playing games of stalk and chase, they begin to hunt with her or on their own.

Cheetahs usually live alone. But females are often seen with their cubs. Males occasionally travel in small groups of two or three animals. The cats communicate by marking tree trunks, bushes, and termite mounds with their waste. By the smell, other cheetahs know that an animal has passed by.

Female cheetahs usually bear three cubs in a litter. Cubs have blue-gray fur on their heads and backs until they are about three months old. To clean their fur, they lick themselves and each other. Often after eating, the cubs and mother groom each other with their long, pink tongues. They close their eyes, lick each other, and purr loudly.

Cheetahs do not roar as lions or tigers do. When alarmed, a cheetah may whine or growl. A cub makes chirping sounds to call its mother.

For thousands of years, cheetahs were kept by royalty as hunting companions. In the 16th century, emperors of India used trained cheetahs to bring down antelopes. But today cheetahs are extinct in India. A few are left in the Middle East, and some remain in Africa. They have little land to roam because the grasslands on which they live have been broken up for farms and ranches. People still hunt the animals for their spotted skins. To help protect cheetahs from hunters, the United States has laws against importing their fur.

CHEETAH

LENGTH OF HEAD AND BODY: 44-59 in (112-150 cm); tail, 23-31 in (58-79 cm)

WEIGHT: 77-143 lb (35-65 kg)

HABITAT AND RANGE: grasslands and open woodlands in Africa and the Middle East

FOOD: antelopes, sheep, and smaller animals

LIFE SPAN: 10 to 12 years in the wild

REPRODUCTION: 1 to 6 young after a pregnancy of 3 months

ORDER: carnivores

▽ *Always alert, cubs and their mother watch for gazelles in the distance, even when resting.*

Chimpanzee

NOISY AND CURIOUS, intelligent and social, the chimpanzee is the wild animal most like a human. Maybe that's why zoo visitors seem to love chimps so much. Chimpanzees—like orangutans, gorillas, and gibbons—are apes. Apes belong to the primate order, a group that also includes lemurs, monkeys, and humans.

Chimps have long arms and short legs. Long black hair covers much of their bodies. Their faces, ears, fingers, and toes are bare. A full-grown male chimp usually measures about 4 feet (122 cm) long and weighs about 100 pounds (45 kg). Females are slightly shorter and lighter. The pygmy chimp is a smaller kind of chimpanzee.

Chimpanzees are found in the dense rain forests, in the open woodlands, and on the broad grasslands of Africa. Chimps spend much of their time on the ground, traveling on all fours. They may

walk upright for a short distance, especially if they are carrying something. They climb into trees to sleep and sometimes to find food.

Chimps are active during the day, searching for food and eating. They feed on fruit, leaves, seeds, buds, bark, stems, and insects. Scientists have also seen chimps catch and eat other small mammals such as young baboons.

Chimpanzees have hands that can grip firmly.

This allows them to pick up and use objects for special purposes. Sometimes the animals use leaves for sponges. They chew on the leaves, which makes them absorbent. Then the animals soak the leaves in water and suck on them to get moisture.

Chimps have been seen using sticks to drive intruders away. Chimps also use a blade of grass or a twig to fish termites or ants out of the ground. A chimp will push a twig into an insect nest, and the

◁ *Chimpanzee finds a comfortable perch on a tree branch. The rain forests and open woodlands of Africa provide places for chimpanzees to play, feed, and sleep. The animals may also travel through grasslands.*

Anything there for me? A female chimp peers at ▷ another female's mouth as she eats. Chimpanzees often share such foods as leaves, fruit, and even small animals.

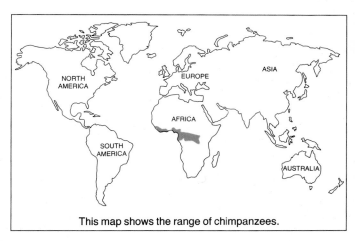

This map shows the range of chimpanzees.

CHIMPANZEE

LENGTH OF HEAD AND BODY: 2-4 ft (61-122 cm)

WEIGHT: 55-110 lb (25-50 kg)

HABITAT AND RANGE: rain forests, open woodlands, and grasslands of Africa

FOOD: fruit, leaves, bark, stems, seeds, buds, insects, and other small animals

LIFE SPAN: 35 to 40 years in the wild

REPRODUCTION: usually 1 young after a pregnancy of about 8¹/₂ months

ORDER: primates

Female chimp cradles her eight-month-old offspring ▷ —a rare set of twins. Mothers nurse their young for several years. At times, other females help baby-sit.

ants or termites will cling to it. The chimp then pulls the twig out and picks the insects off with its lips.

Chimps live in communities of about fifty animals that share the same area. Within these large communities, chimps form smaller groups of three to six animals. The chimps in these groups travel together for a while. But they do not always stay together. Groups are always changing as chimps choose new companions. Most groups are a mixture

▽ *Napping in a day nest in the trees, a chimp stretches out lazily. At night, the chimp will build another sleeping nest. It bends branches and twigs into a comfortable, leafy pad in a tree fork.*

of males, females, and young. At times, however, only females and young remain together. At other times, a chimpanzee may travel alone.

When there is a large amount of food, chimps gather to have a feast. They bark loudly to announce the find. When other chimps hear the call, they rush over to join in the eating. Chimps beg each other for some food by holding out their hands with the palms up.

Chimpanzees use a complicated system of sounds to communicate with each other. For example, a loud call like "wraaaa" warns of something unusual or disturbing. The calls can be heard 2 miles (3 km) away. To express contentment, a chimp grunts softly.

Touch is also important in the lives of chimpanzees. A nervous animal will reach out to touch another chimp. Chimps may kiss when they meet, and they also hold hands. An adult chimp sometimes has a special companion. The two chimps spend time together and comfort each other.

Another way a chimp communicates is by the expression on its face. When a chimpanzee bares its teeth, it lets others know that it is excited or frightened. If the animal grins with its lips covering its teeth, it means that it is in a friendly mood. If it puckers up its lips and looks as if it's about to give someone a big, smacking kiss, the chimp is worried. Another expression—lips pressed together—means the chimp may be about to charge or to attack.

Female chimps give their offspring a great deal of care. The young are born after an eight-and-a-half-month pregnancy. During the first months of a newborn's life, its mother carries it everywhere. The female cradles the tiny chimp carefully as it clings to

◁ *Like a small jockey, a two-year-old chimp rides on its mother's back. The youngster joins in her hooting call. Copying the behavior of an adult animal forms an important part of a chimp's education.*

Can you find the chimp hidden in the trees? Only his ▷ *head shows among the leaves. In the small picture, a mother chimp eats as her offspring clings to her belly. She uses her long, strong arm to hang from a branch. Chimps sometimes use their arms and hands to swing short distances in the trees.*

Chimpanzee

▽ *Young male chimpanzee picks through his mother's hair as she grooms her daughter. Grooming helps keep a chimp's skin and hair clean. It also provides important social contact for chimpanzees in a group.*

△ *Using a blade of grass, a chimp fishes into a termite mound in search of food. The insects inside cling to the grass. Then the chimp pulls the grass out and gobbles up the termites. Chimps also use other objects found in the forests. They may use sticks as weapons. Or they may chew leaves and use them as sponges to soak up water.*

◁ *Two adult chimps open their mouths wide during a series of pants and hoots. Silent, a young chimp clings to its mother. Chimps sometimes pant-hoot when they find food. Chimps communicate many things by their calls. A soft bark means a mild warning. During grooming sessions, animals may grunt as they pick at one another's hair and skin.*

the hair on her belly. At about five months, the young chimp begins to ride on its mother's back. It perches there like a little jockey. Chimpanzee mothers play with their young, grooming and tickling them and sharing their sleeping nests with them at night. Female chimps help each other with baby-sitting chores. Older females will often look after their younger sisters and brothers.

Like humans, chimpanzees grow up slowly. When about nine years old, they begin adolescence. By the time they are 12 years old, they may have offspring of their own.

As young chimps grow and become better able to care for themselves, they play with other young chimps. Older chimpanzees in the community usually are patient with the energetic youngsters. They allow them to do pretty much as they please. And the young chimps do! They climb trees, wrestle with each other and their elders, and play with sticks, food, and other objects.

Play helps young chimpanzees learn about their world. By wrestling, they learn how strong they are and what they can do. They learn which branches are big enough to hold them. Young chimps also learn by trying to do what adults do. They make leafy sleeping nests. And they practice making the expressions their elders make and the calls they give. All these activities are important to a young chimp's development. They help the chimp learn to take care of itself as an adult.

The intelligence of the chimpanzee has enabled scientists to teach captive animals many things. By studying how these primates react, experts can find out more about the learning process.

In the wild, the number of chimpanzees has become smaller. The wilderness in which the animals live is gradually disappearing. In recent years, scientists have studied chimps in Africa. What they have learned may make the difference between survival and extinction for chimpanzees.

Chinchilla

(*say* chin-CHILL-uh)

COLD WINDS whip through the high Andes of South America. Few plants cover the rugged mountains. The wild chinchilla moves easily among the rocks in search of food. The cold does not bother the small rodent because it is covered with thick fur.

The chinchilla's head and body measure only about 10 inches (25 cm) long, and its fur is about 1 inch (3 cm) long. The soft, fine hairs range in color from bluish gray to brownish gray. Each hair may be tipped with black.

Wild chinchillas feed in the morning and in the evening, coming out of hiding places among the rocks to eat bark, grasses, and herbs. Sitting up, they hold their food in their forepaws as they gnaw on it. They do not drink much water. Chinchillas get most of the moisture they need from plants.

Chinchillas are born in litters of one to four young. They have fur at birth.

Long ago, Indians of South America used the chinchilla's soft fur for blankets and clothes. Today people still prize the fur, and wild chinchillas have been hunted until few are left. Most chinchillas now are raised on farms in several parts of the world. Their fur is used in making coats and jackets.

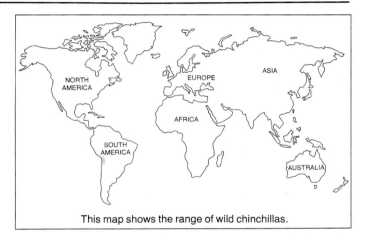

This map shows the range of wild chinchillas.

▽ *Chinchilla nibbles a plant with its sharp teeth. Like all rodents, a chinchilla must gnaw on hard substances to wear down its front teeth, which never stop growing.*

CHINCHILLA

LENGTH OF HEAD AND BODY: 9-15 in (23-38 cm); tail, 3-6 in (8-15 cm)

WEIGHT: 18-28 oz (510-794 g)

HABITAT AND RANGE: parts of the high Andes in South America; chinchillas are raised for their fur in several parts of the world

FOOD: bark, grasses, and herbs

LIFE SPAN: 15 to 20 years in captivity

REPRODUCTION: 1 to 4 young after a pregnancy of about 4 months

ORDER: rodents

Chipmunk

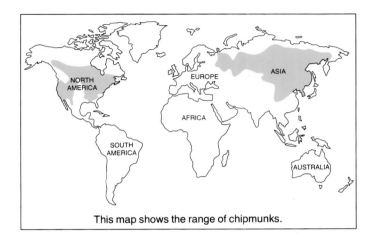

This map shows the range of chipmunks.

CHIPMUNK

LENGTH OF HEAD AND BODY: 4-7 in (10-18 cm); tail, 3-5 in (8-13 cm)

WEIGHT: 1-5 oz (28-142 g)

HABITAT AND RANGE: forests, open woodlands, and brushy areas in North America and Asia

FOOD: nuts, berries, seeds, fruit, grain, and insects

LIFE SPAN: about 2 or 3 years in the wild

REPRODUCTION: 2 to 8 young after a pregnancy of about 1 month

ORDER: rodents

Least chipmunk: 4 in (10 cm) long; tail, 3 in (8 cm)

△ *Least chipmunk rests on a tree stump on a spring day. Smallest of about 17 kinds of chipmunks, the least chipmunk lives in parts of North America.*

SITTING UP, TAIL TWITCHING, forepaws clasped to its chest, a chipmunk sings to its neighbors. "Chip, chip," pipes the little striped rodent. When it spies an enemy—a weasel, a fox, a hawk, or a snake—a chipmunk calls out a sudden warning and dashes for cover.

There are about 17 kinds of chipmunks. They range throughout North America and Asia. They are at home in forests, in open woodlands, and even in city parks. Some dig burrows. Others build nests in bushes or logs. You have probably seen one of these grayish or brownish animals on the ground or on a branch.

Chipmunks spend the summer eating plants and insects. Then, in early fall, they begin to gather nuts and seeds for winter. They tuck the food under rocks and logs or inside their burrows. As cold weather sets in, most chipmunks move underground to sleep. From time to time, they wake up and eat the food they have stored.

When spring arrives, chipmunks emerge from their burrows and find mates. Females bear two to eight young about four weeks later. For two months, the parents care for their offspring. Then the young chipmunks begin putting away their own food for the winter ahead.

Cheek pouches packed with food, a watchful eastern chipmunk (left) pauses on a limb. It can stuff several acorns at a time into its mouth. Gathering nuts and seeds takes up most of a chipmunk's time. At right, another eastern chipmunk emerges from its burrow.

Eastern chipmunk: 7 in (18 cm) long; tail, 4 in (10 cm)

Civet

Fanaloka: 16 in (41 cm) long; tail, 8 in (20 cm)

Fanaloka, a small civet of Madagascar, balances on a rock. This rare animal usually lives in forests.

FORESTS, GRASSLANDS, AND MARSHES—lean, long-tailed civets roam them all. Civets live in a variety of habitats and are known by many names. Genets, linsangs, and binturongs all are kinds of civets. You can read about them under their own headings in this book.

Civets live in parts of Asia, in central and southern Africa, and on the island of Madagascar. They make their homes in trees, in piles of brush, among roots, and between rocks.

Palm civets spend most of their time in trees. They sleep curled up on a branch or in a tangle of tree limbs. Palm civets eat small animals as well as fruit they find among the leaves and on the ground.

The African civet stays on the ground. It sleeps in a grassy bed or in a rocky shelter. When it wakes, it pads softly through the underbrush in search of such food as insects, rodents, birds, and fruit.

The otter civet is adapted, or suited, to life in the water. This strong swimmer gets its food by catching frogs and fish in rivers of southern Asia.

Most civets sleep through the tropical heat of the day and awake to hunt at night. Their striped and spotted fur blends in well with their shadowy surroundings. Though they usually live alone, civets have a way to communicate with each other. They use their sense of smell.

Civets have many scent glands in their bodies. As a civet travels, it makes scent marks along its path. These marks let others know that a civet has passed by. By sniffing, a civet can tell whether the civet that left the mark is looking for a mate.

Some civets have glands under their tails that produce a strong-smelling oil. For centuries, people have used this oil from captured civets to make perfume. People have been familiar with some kinds of civets for a long time. But not much is known about these animals in the wild.

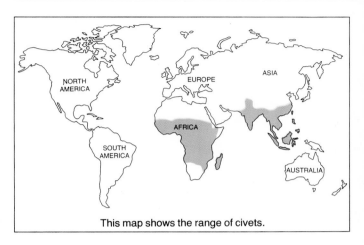

This map shows the range of civets.

CIVET

LENGTH OF HEAD AND BODY: 16-35 in (41-89 cm); tail, 7-26 in (18-66 cm)

WEIGHT: 2-44 lb (1-20 kg)

HABITAT AND RANGE: forests, woodlands, and grasslands of Madagascar, southern and southeastern Asia, and central and southern Africa

FOOD: fruit and small animals, including rodents, birds, reptiles, shellfish, and insects

LIFE SPAN: as long as 15 years in captivity

REPRODUCTION: 1 to 4 young after a pregnancy of 2 or 3 months

ORDER: carnivores

Small-toothed palm civet: 18 in (46 cm) long; tail, 21 in (53 cm)

△ *Small-toothed palm civet clutches an egg in its front paws. Although palm civets eat mostly fruit, they sometimes vary their diet by raiding birds' nests. The animal's grayish coloring blends well with its surroundings in the forests of Asia.*

▽ *Nose to the ground, an African civet follows a scent. Like some other civets, this animal produces oil in a gland under its tail. By pressing its rump against trees, bushes, and rocks, it makes a scent mark.*

African civet: 33 in (84 cm) long; tail, 18 in (46 cm)

155

Coati

THEY SNORT AND SNUFFLE along the ground, looking for something to eat. As coatis wander through a tropical forest, they rustle through the leaves. They grunt softly to one another.

A member of the raccoon family, the coati is a very social animal. About four to twenty female coatis travel with their young in a group called a band. The band spends most of the day searching for food. From time to time, coatis stop moving around. As they rest, they groom each other by nibbling at one another's fur. At night, they sleep in the trees. Male coatis live alone, except during the mating season.

Yawning after a nap, a sleepy coati bumps its ▷ *nose on a tree branch. Because of the coati's long snout, people sometimes call the animal the hog-nosed coon.*

▽ *Bushy brown-and-white tails wave in the air as a group of ring-tailed coatis in Brazil looks for food. Because of the animals' tails, people sometimes mistake coatis for monkeys. Coatis really belong to the raccoon family.*

Coati: 25 in (64 cm) long; tail, 25 in (64 cm)

Ring-tailed coati: 25 in (64 cm) long; tail, 25 in (64 cm)

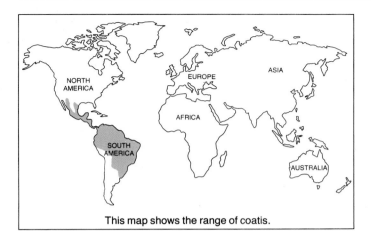

This map shows the range of coatis.

▽ *Extending its long tail for balance, a coati climbs a tree in Arizona. Coatis feed on fruit and insects in trees. To get the food, they may hang upside down on branches.*

COATI

LENGTH OF HEAD AND BODY: **13-27 in (33-69 cm); tail, 13-27 in (33-69 cm)**

WEIGHT: **7-15 lb (3-7 kg)**

HABITAT AND RANGE: **dense forests, grasslands, and brushy areas from the southwestern United States through northern Argentina**

FOOD: **insects, snails, small reptiles, rodents, fruit, and nuts**

LIFE SPAN: **14 years in captivity**

REPRODUCTION: **2 to 6 young after a pregnancy of about 2½ months**

ORDER: **carnivores**

Females leave their bands to bear young—usually two to six. They have their young in nests that they build in trees. In five or six weeks, they return to the group, bringing their offspring with them.

Young coatis join their mothers in search of food as soon as they leave the nest. But they play much of the time, wrestling or chasing one another among the trees. The young will be full grown when they are two years old.

The coati is sometimes called the hog-nosed coon because of its long snout. Its nose is very sensitive. Coatis find food by sniffing until they detect an animal or a piece of fruit. When a coati smells prey underground, it uses strong claws to dig it out.

Each band of females and young has a home area where it searches for food. Each male coati has its own territory. At certain times of the year, coatis travel more widely to find food. When fruit is in season, coatis might make a special trip to a place where many fruit trees grow.

Most coatis are found in dense, wet forests from Mexico into South America. Since the beginning of the century, they have moved into the southwestern United States, even where there are few trees. The coatis that live in the Southwest often sleep in caves and rock piles, instead of on branches.

▽ *Coati drinks from a small pool in Arizona. Most coatis live in wet forests in Central and South America, but their range has spread north to drier areas.*

Colobus

The colobus is a kind of monkey. Read about monkeys on page 376.

Cougar

Cougar is another name for mountain lion. Read about it and other wild cats on page 126.

Cow

Texas Longhorn: 58 in (147 cm) tall at the shoulder

This map shows the range of wild cows.

◁ *On a grassy prairie, a Texas Longhorn bull chews his cud. His horns measure about 4 feet (122 cm) from tip to tip. During the 1800s, cowboys herded millions of Longhorns on cattle drives north from Texas.*

WE SEE THEM—and hear them mooing—when we visit a farm. Some cows are brown and white, some are tan, some are red, and some are black and white. Others are pure black or pure white with large humps on their shoulders. Cows are some of the most familiar farm animals in the world. We also use the word "cow" to refer to a female. Males are bulls, and the young are calves. Cows in general are often called cattle.

Cows have been domesticated—tamed and raised by people—for thousands of years. Cows are native to Europe, Asia, and Africa. They are now found in many other parts of the world. In some places in Africa, a person's wealth may be judged by the number of cows he owns.

All cows have long, tufted tails. They use them as switches to help keep pesky insects off their backs and sides. You probably have seen cows standing in fields, swinging their tails back and forth to keep flies away. As they stand there, they calmly chew wads of food called cuds.

A cow has a special stomach with four compartments that digest the huge amount of grass the animal eats. A cow swallows its food after chewing it only a little. In the first and second parts of its stomach, the food combines with liquid and forms into cuds. Later, the cow brings up a cud and chews it more completely. Then it swallows the cud, which ends up in the third and fourth parts of the stomach to be further digested. Some other animals that chew cuds are antelopes, deer, goats, and sheep.

On farms around the world, cows are raised for two important kinds of food: milk and meat. Milk cows are called dairy cows. In many parts of the world today, dairy farms have modern milking equipment. Machines squeeze the milk from a cow's udder—the baglike part of her body where the milk collects. But cows are still milked by hand in many places. Someone will sit beside the animal and gently squeeze the milk into a pail.

The large, black-and-white Holstein-Friesians (say HOLE-steen FREE-zhunz) are the most common

Hereford: 52 in (132 cm) tall at the shoulder

Hereford cows and calves, raised for beef, huddle against November winds in a Montana pasture.

Highland: 54 in (137 cm) tall at the shoulder

Holstein-Friesian: 58 in (147 cm) tall at the shoulder

△ *Coarse, shaggy coat of a Highland bull protects it from harsh weather in Scotland. People have raised these hardy animals for centuries for their meat.*

△ *Newborn Holstein-Friesian calf stands for the first time. Its mother watches her unsteady offspring. Cows usually bear one young every year.*

159

dairy cows in the United States. Some other widespread dairy breeds are brown Jerseys and tan-and-white Guernseys (say GURN-zeez). Both Jerseys and Guernseys are famous for their creamy yellow milk, which is rich in butterfat.

Herefords (say HER-ferdz) are a popular beef breed. You can recognize them by their white faces and reddish coats. Sometimes their coats are quite curly. The Hereford breed came from England originally, but these cattle now are found on farms and on ranches all over the United States. Aberdeen Angus, beef cattle originally from Scotland, are thickset animals with black bodies. Unlike many other breeds, Aberdeen Angus cattle have no horns.

Besides providing milk and meat, cows give us hides for leather as well as substances used in medicine, soap, and glue. In some parts of the world, domestic cattle are still used to pull carts and plows. Cows are useful—most of the time. According to a legend, the Great Chicago Fire of 1871 started when Mrs. O'Leary's cow kicked over a lantern in a barn.

Most cows give birth to a single calf after a pregnancy of about nine months. Calves may nurse for as long as eight months. During this time, they also eat grass.

Modern breeds of cows can be traced back to wild European cattle called aurochs (say ow-rocks). Aurochs are now extinct. But we know what the animals looked like. Thousands of years ago, people painted pictures of aurochs on the walls of caves.

Several kinds of wild cattle—sometimes known as oxen—live in Asia. Bantengs (say BON-tengz) roam the forests of Southeast Asia. They have brown coats and white rump patches. Marks that look like white stockings cover their legs. Bantengs live in herds and are often active at dawn, at dusk, and sometimes even at night. Some have been domesticated and live on farms.

Gaur, large dark brown cattle, live in forests in

▽ *Herd of gaur—females and young—graze in a grassy clearing in India. The largest wild cattle, gaur live in forested hills. They usually move around and feed between dusk and dawn.*

Gaur: 72 in (183 cm) tall at the shoulder

Zebu: 56 in (142 cm) tall at the shoulder

◁ *Zebus wander through a meadow in Brazil. These humpbacked cattle, originally from India, survive easily in hot climates. Ranchers often breed them with other cattle to produce even hardier animals.*

COW

HEIGHT: 46-75 in (117-191 cm) at the shoulder

WEIGHT: 325-3,000 lb (147-1,361 kg)

HABITAT AND RANGE: grassy plains, open and dense woodlands, rain forests, and hilly areas in India, Nepal, Burma, and Southeast Asia; domesticated cows live in many parts of the world

FOOD: grasses, herbs, leaves, twigs, and bamboo shoots

LIFE SPAN: 12 to 26 years in captivity, depending on species

REPRODUCTION: usually 1 young after a pregnancy of about 9 months

ORDER: artiodactyls

Banteng bull in Java (below) swishes his tufted tail and tosses his slender horns. In the wild, bantengs roam in herds of as many as 25 animals. Two Ankole cattle in Africa chew their cuds (below, left). A female's horns can measure 5 feet (152 cm) across. But a bull's horns do not grow as big. Can you recognize the female here?

India and Southeast Asia. They often feed in the evening and in the early morning on grasses, herbs, and leaves. They are the largest of all wild cattle. Like bantengs, gaur have stockinglike white marks on their lower legs. Their horns curve in a half-moon shape toward each other. In spite of their size, gaur can move quickly in times of danger. They make a snorting sound as they speed away.

In India, people of the Hindu religion consider cows sacred. They do not kill or eat the animals. Indian cows—or zebus (say ZEE-booz)—are often grayish white, though some zebus may be black or red. The zebu has a hump on its shoulders. A fold of skin, called a dewlap, hangs down from its neck. Sometimes the dewlap almost reaches the ground. Zebus give milk and are used to pull plows. Some zebus wander freely in cities and towns. They often cause trouble, stopping traffic and eating food from open-air markets.

Ankole: 52 in (132 cm) tall at the shoulder

Banteng: 59 in (150 cm) tall at the shoulder

Coyote

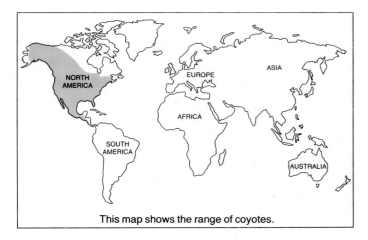

This map shows the range of coyotes.

◁ *Alone and watchful, a coyote sits among golden grasses in Yellowstone National Park. People have nicknamed this sly animal "prairie wolf" after its larger relative, the wolf.*

▽ *Lone coyote sings to the sky. By sending its call across the prairie, it lets other coyotes know its whereabouts. In the evening, many animals may join the chorus.*

IN SOME INDIAN STORIES, the trickiest creature of all is Coyote. And in real life, coyotes *are* clever animals. They quickly take advantage of changes around them. Once coyotes lived mainly on the prairies and in the deserts of North America. But as people settled across the land, coyotes learned to survive in mountains and in forests, too. Now they roam throughout much of the continent.

The coyote's tan coat is usually mixed with hairs of rusty brown and gray. This helps the coyote hide in grasses, among rocks, or in underbrush.

As it hunts, the coyote uses its sharp eyesight, keen hearing, and sensitive nose. It trots long distances in search of prey.

Coyotes will eat almost anything. In the summer, they usually feed on mice, rabbits, and insects. Coyotes also catch fish and frogs. They pounce on snakes and lizards. They even feed on grasses, nuts, and fruit—including watermelon! Coyotes eat dead elk and deer that they find—especially in winter.

Coyotes and badgers sometimes help each other when they hunt. Chased by a coyote, a rabbit or a mouse may run into a hole. There a badger can dig it out. At other times, a badger may dig for a ground squirrel, only to have it pop out of another entrance and escape. When the ground squirrel emerges, a waiting coyote can grab it.

△ *Coyote pounces on its prey—perhaps a mouse scurrying through the grass. Coyotes use their keen senses when they hunt for food.*

163

Coypu

WHEN IT HEARS A NOISE, the shy coypu perks its small ears and wiggles its long whiskers. Then this furry, groundhog-size rodent runs to the water's edge and jumps in. Clumsy on land, the coypu moves gracefully in the water. Paddling with its webbed hind feet, it swims quickly to safety.

The coypu digs a burrow in the soft earth of a riverbank. Or it may build a nest of reeds in a marsh or along a lakeshore. In these wet places, the coypu searches for plants, mussels, and snails. With its forepaw it skims the water for food or reaches underwater to yank out a reed or a root.

Coypus live in pairs or in large colonies. Females give birth to two or three litters every year. Usually a litter has five to seven young. When the newborn are only five days old, they are able to survive on their own. Usually, though, they stay with their mothers for six to eight weeks.

Trappers have long hunted the coypu for its thick, valuable fur. Beneath the coarse, reddish

brown outer hair is an undercoat of brown or dark gray. This soft fur, called nutria (say NYU-tree-uh), is sometimes made into winter jackets and overcoats. Nutria is also another name for the coypu.

Coypus originally were found only in southern South America. Today they are raised in many parts of the world for their fur. Sometimes coypus escape from the fur farms, and many of the animals now live wild in North America. In some areas, they are pests because they destroy crops.

COYPU

LENGTH OF HEAD AND BODY: **17-25 in (43-64 cm); tail, 10-16 in (25-41 cm)**

WEIGHT: **15-22 lb (7-10 kg)**

HABITAT AND RANGE: **near rivers and lakes in North and South America**

FOOD: **water plants, mussels, and snails**

LIFE SPAN: **about 6 years in captivity**

REPRODUCTION: **usually 5 to 7 young after a pregnancy of about 4 months**

ORDER: **rodents**

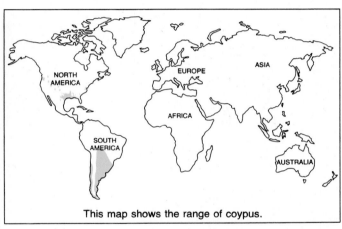

This map shows the range of coypus.

◁ *Curving its long tail under its chin, a coypu suns itself on a post. It feeds on the plants floating nearby.*

▽ *Orange teeth gleaming in the sun, a coypu paddles through the water with its webbed hind feet. The animal can dive underwater and stay for five minutes.*

Spotted cuscus: 20 in (51 cm) long; tail, 17 in (43 cm)

Cuscus

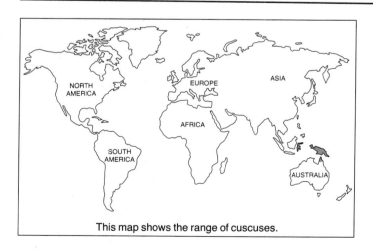

This map shows the range of cuscuses.

ALL DAY LONG, the furry cuscus sits in a treetop overlooking its dense forest home. The cuscus eats and often sleeps in a sitting position. In fact, this animal spends so much time sitting that it often rubs a bare place on its rump!

A good climber, the cuscus hardly ever comes down from the trees. To help keep a firm grip on the branches, the cuscus uses its hind paws as well as its front paws. The animal wraps its strong, thick tail around tree limbs for more security.

During the night, the cuscus moves along tree branches looking for food. Occasionally, it quietly creeps up on a lizard or a bird and grabs it with its front paws. This adds variety to the leaves, fruit, and insects that it usually eats.

Cuscuses rarely hurry. They are usually safe in the treetops. The animals have few enemies where they live on the northeastern tip of Australia, in New Guinea, and on nearby islands. If another animal does threaten or annoy a cuscus, the animal will strike with its front paws. It barks and snarls sharply, frightening away the intruder.

Some people consider the cuscus one of the most colorful mammals in the world. Many cuscuses have bright yellow noses and bulging yellow, orange, or red eyes. Their fur comes in many colors—from white or yellow, to black or grayish green. Patterns often decorate their woolly coats.

◁ *High-wire artist, a spotted cuscus creeps along a branch by gripping with all four paws. For extra support, the cuscus can curl its tail around a tree limb.*

▽ *Big eyes stare out from the colorful face of a spotted cuscus. Keen sight helps this animal move about the trees at night in search of lizards, birds, insects, and fruit.*

△ *Quiet during the day, a "spotless" spotted cuscus nestles in the branches. Cuscuses wear coats of many colors and patterns. An animal's fur may change color several times as it grows older.*

The cuscus's fur often grows so thick that it covers the animal's small ears. The cuscus grooms its fur by combing it with its claws.

The cuscus is about the size of a monkey, and some people mistake it for one. However, cuscuses belong to the group of pouched mammals called marsupials (say mar-soo-pea-ulz). Marsupials give birth to very small, underdeveloped young. These offspring stay in their mother's protective pouch for several months, until they are bigger. A female cuscus bears one or two young each year. Find out about other marsupials by reading the entries on kangaroos on page 310 and koalas on page 318.

CUSCUS

LENGTH OF HEAD AND BODY: **13-26 in (33-66 cm); tail, 10-25 in (25-64 cm)**

WEIGHT: **as much as 11 lb (5 kg)**

HABITAT AND RANGE: **forests of northeastern Australia, New Guinea, and neighboring islands**

FOOD: **leaves, fruit, insects, lizards, small birds, and eggs**

LIFE SPAN: **3 to 11 years in captivity, depending on species**

REPRODUCTION: **1 or 2 young after a pregnancy of about 2 weeks**

ORDER: **marsupials**

D

Powerful hind legs of a white-tailed deer buck swing forward in a bounding stride. White-tailed deer can run for several

Deer

EVERY SPRING, new sets of antlers begin to sprout from the heads of male deer in many parts of the world. Antlers are not the same as horns. Antlers are bones that develop and usually fall off each year. Horns grow out of the skin, just as hair and fingernails do. They never stop growing.

When antlers first begin to develop, they look like two bumps on the top of a deer's head. They grow quickly and branch out. But the antlers remain soft and tender for the first few months, until they reach their full size. Velvet, a layer of skin and fine hairs, covers the antlers. Late in the summer, when the antlers have stopped growing, the velvet dries up. The male deer rubs it off on trees and bushes, revealing the hard, sharp points of the antlers.

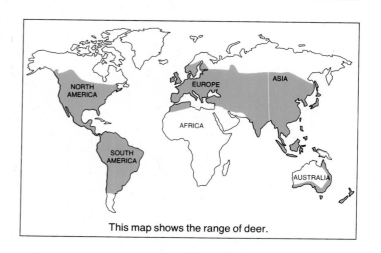

This map shows the range of deer.

White-tailed deer: 40 in (102 cm) tall at the shoulder

miles at speeds of 30 miles (48 km) an hour.

△ *Shielded by its mother, a young white-tailed deer watches wood ducks in Florida. The young stays with its mother for one or two years. Deer often feed on plants near ponds and streams.*

◁ *Spotted coat of a six-week-old white-tailed deer blends well with grass and bushes. Because the newborn have no scent, even enemies with keen senses of smell cannot easily find them.*

171

Deer

Of the many species, or kinds, of deer, only two do not have antlers. Musk deer and Chinese water deer have tusks, or long teeth, instead. Both of these kinds live in Asia. Other deer, the Asian muntjac (say MUNT-jack) and the South American guemal (say GWAY-mul) have both tusks and antlers. The female deer of most species do not develop antlers.

A male deer, often called a buck, carries his

Red deer: 45 in (114 cm) tall at the shoulder

△ Antlers clash as young male red deer in Scotland test their strength. These close relatives of North American elk live in many parts of Europe.

Antlers come in a ▷ variety of shapes. A sambar deer of India (right) has thin, branching antlers. But the antlers of a fallow deer (far right) are shaped like hands with the fingers spread. Older males usually have larger antlers.

172

Sambar deer: 54 in (137 cm) tall at the shoulder

Fallow deer: 37 in (94 cm) tall at the shoulder

antlers for several months. In winter, after the mating season, the bony branches fall off. First one antler and then the other drops to the ground. The buck has no antlers until the following year. Smaller animals like mice, squirrels, and porcupines gnaw on the old antlers. The bones are rich in the minerals that these animals need.

In the mating season, bucks strut and show off their antlers. They use them to keep rivals away and to attract the attention of females—often called does. Two bucks threaten each other by snorting and by shaking their heads. If the bucks spar, they shove and wrestle with their antlers and tear up the earth with their hooves. Their antlers may crash together, but the sharp tips seldom reach the bucks' bodies. Such a match may last for two hours, until one buck gives up and dashes off. Usually neither deer is seriously hurt. The winner and the doe stay together for several days. Then he leaves her to find another mate.

Most female deer give birth to one or two young each spring, six to nine months after mating. The fawns, or offspring, of many kinds of deer have spotted coats. But the patterns disappear after a few months, when the fawn grows its winter coat. In a few kinds of deer—the fallow deer of Europe and the axis deer, hog deer, and sika deer of Asia—adults have spots, too. *(Continued on page 176)*

Every spring, a male white-tailed deer begins to grow new antlers. First, knobs appear (1). By late summer, antlers have formed (2) under a cover of skin called velvet. The velvet peels off in strips (3), revealing sharp points. In winter, the antlers fall off, usually one at a time (4). They leave small wounds that heal quickly.

Short, hooked antlers of an Asian muntjac (below, left) sprout from bony ridges on its face. Instead of antlers, a male musk deer of Asia (below, center) has sharp tusks that measure 2 inches (5 cm) long. A brocket deer in Mexico (below, right) has short, spiky antlers. Brocket deer can grow antlers at any time of year.

Muntjac: 21 in (53 cm) tall at the shoulder

Musk deer: 22 in (56 cm) tall at the shoulder

Brocket deer: 26 in (66 cm) tall at the shoulder

Mule deer: 39 in (99 cm) tall at the shoulder

△ Pawing the soft snow, a mule deer buck searches for food. Large ears give the mule deer its name. It lives in mountains, plains, and deserts of the western United States. Bucks usually roam alone or in small groups.

Golden coats of swamp ▷ deer gleam in the sunshine after an April shower in India. A group of male and female adults and young feeds on lush grasses that grow in the rainy season. The hooves of these deer spread out to support their weight on soft, wet ground.

Swamp deer: 45 in (114 cm) tall at the shoulder

Guemal fawn copies its mother's alert pose. Guemals live in the rugged Andes of ▷ South America, from Ecuador through Chile. In winter, Chilean guemals leave the mountains to find shelter in the forests below.

Guemal: 33 in (84 cm) tall at the shoulder

Hog deer: 28 in (71 cm) tall at the shoulder

△ *Heavyset, short-legged body of a hog deer gives the animal its name. The hog deer does not run gracefully as most deer do. Instead it dashes through tall grass like a wild pig. Called the paddy-field deer in Sri Lanka, it feeds on growing rice.*

175

The spotted coats of the fawns blend with forests and grasslands and make the young hard to see. Newborn fawns are also protected from danger because they have no scent. Even a dog, with its sensitive nose, may not smell a nearby fawn.

When the fawn is strong enough, its mother takes it to join a group of other does and their young. Fawns play together. They stand on their hind legs and box with their forelegs. Or they butt and chase each other. Their safety from enemies may depend on the lessons they learn during such play. Many animals prey on deer. In North America, the deer's natural enemies include coyotes, wolves, bears, and mountain lions. Elsewhere, deer are hunted by big cats like tigers, leopards, and jaguars.

Deer feed early in the morning and again in the evening. They tear off bites of grass, bark, leaves, and twigs and then swallow the food almost whole. After eating, when a deer lies hidden, it brings up a wad of partly digested food—called a cud—from its stomach. The deer chews the cud thoroughly, swallows it, and digests it completely.

Because they have so many enemies, deer must be alert. They rely mostly on their keen senses of sight, smell, and hearing to warn them of enemies. When danger first threatens, deer may freeze. Then they run away. White-tailed deer can gallop for about 4 miles (6 km) at 30 miles (48 km) an hour. They usually run much shorter distances, however. They try to find a hiding place in a grove of trees or just over a hill from danger.

Today deer are found on all continents except Antarctica. The smallest deer, the pudu, is the size of a raccoon. It weighs about 20 pounds (9 kg). The moose, the largest member of the deer family, can measure 7 feet (213 cm) tall at the shoulder. It can weigh 1,800 pounds (816 kg).

In North America, most deer are white-tailed deer. About 12 million of them live in the United States, making them the most common of the country's large mammals. White-tailed deer may grow $4^1/_2$ feet (137 cm) tall at the shoulder and weigh as much as 300 pounds (136 kg). West of the Mississippi River, the most common deer are mule deer. They get their name from their long, mulelike ears.

For centuries, people have hunted white-tailed deer for food, for sport, and for their skins. The skins were used for jackets and other clothing and for moccasins. Many years ago, people on the frontier even used deer hides, known as buckskins, as a kind of money. That's why people sometimes call a dollar bill a "buck."

Find out about other members of the deer family by reading about caribou on page 120, elk on page 198, and moose on page 392.

Pudu: 15 in (38 cm) tall at the shoulder

△ *Pudu doe picks her way through the grass. The smallest of all deer, the raccoon-size pudu makes its home in forests and mountains in South America.*

DEER

HEIGHT: **15 in-7 ft (38-213 cm) tall at the shoulder**

WEIGHT: **20-1,800 lb (9-816 kg)**

HABITAT AND RANGE: **woodlands, mountains, forests, grasslands, and deserts in Europe, Asia, northern Africa, the Middle East, New Zealand, North and South America, Australia, and some Pacific islands**

FOOD: **grasses, bark, twigs, and leaves**

LIFE SPAN: **8 to 16 years in the wild**

REPRODUCTION: **1 to 3 young after a pregnancy of 5 to 10 months, depending on species**

ORDER: **artiodactyls**

Like mirror images, male sika deer in Japan rear up ▷ *and prepare to box. With their antlers still soft and sensitive in velvet, the males settle disputes by fighting with their forelegs.*

Sika deer: 35 in (89 cm) tall at the shoulder

Dik-dik

The dik-dik is a kind of antelope. Read about antelopes on page 52.

Dingo

(*say* DING-go)

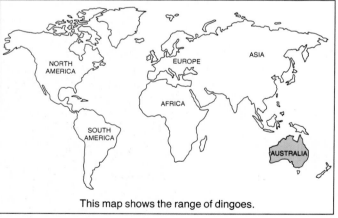

This map shows the range of dingoes.

◁ *Male dingo, a wild dog of Australia, watches the grasslands. Dingoes sometimes howl like wolves.*

DENSE FORESTS and open, dry plains provide homes for the dingo of Australia. Dingoes are members of the dog family. Scientists know little about their origin. They may have arrived as tame dogs with people who came to Australia about 8,000 years ago. Then they may gradually have become wild again.

Dingoes hunt at night, alone or in family groups. They eat rats, rabbits, lizards, birds, and kangaroos. Sometimes they prey on farm animals.

Dingoes make their dens in underground burrows or in hollow logs. There a pair has a litter of four or five pups each year. Pups may stay with their parents for two years and help raise the next litter.

Sometimes Aboriginals, the native people of Australia, train dingoes as hunting dogs. They capture and raise wild pups. Dingoes once even served as living blankets. A chilly night was a "three-dog night." A really cold one was a "six-dog night."

DINGO

LENGTH OF HEAD AND BODY: 46-49 in (117-124 cm); tail, 12-13 in (30-33 cm)

WEIGHT: 22-33 lb (10-15 kg)

HABITAT AND RANGE: forests, open plains, deserts, and rocky mountains in Australia

FOOD: rabbits, rodents, lizards, birds, fruit, marsupials, sheep, and cattle

REPRODUCTION: 4 or 5 pups a year after a pregnancy of 2 months

LIFE SPAN: unknown

ORDER: carnivores

Dog

"MAN'S BEST FRIEND," the dog was probably the first animal ever to be tamed. Today domestic, or tame, dogs are our companions at work and at play, in the city and in the countryside, in North America and all over the world.

All domestic dogs make up only one of about 35 species, or kinds, of dogs. All other dogs live in the wild.

Wild dogs, like wild cats, are carnivores (say CAR-nuh-vorz). That means they are mainly meat-eating hunters. But dogs and cats hunt in different ways. Dogs usually run down their prey. Most cats lie in wait or creep up on their victims. Cats almost always hunt alone. Dogs may hunt alone or in pairs to find small prey. Larger dogs can bring down animals much bigger than they are by hunting together in packs. Smaller wild dogs eat small mammals, insects, fruit, and birds.

Long legs and deep chests give most dogs the speed and the strength they need to catch swift prey. Pads on the bottoms of their paws cushion their feet as they run.

When they have chased their prey to exhaustion, wild dogs attack and kill with powerful jaws. Front teeth bite flesh. Back teeth cut the meat into large chunks. Dogs do little real chewing. They simply swallow their food.

Wild dogs have keen senses. Most find prey by sniffing the air or the ground. A dog's sense of smell is much better than that of a human being. On open plains, sharp-eyed dogs can spot the movement of distant prey. In brushy areas, their ears catch sounds of small, scurrying animals.

Some wild dogs mate for life. A female has one

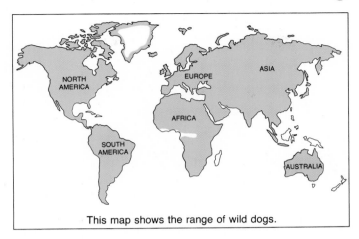

This map shows the range of wild dogs.

DOG

LENGTH OF HEAD AND BODY: 14-57 in (36-145 cm); tail, 5-20 in (13-51 cm); domestic dogs are measured in height at the shoulder—6-33 in (15-84 cm)

WEIGHT: 3-175 lb (1-79 kg)

HABITAT AND RANGE: all kinds of habitats almost everywhere in the world; domestic dogs live in many parts of the world

FOOD: animals and plants

LIFE SPAN: 12 to 20 years in captivity, depending on species

REPRODUCTION: 2 to 12 pups after a pregnancy of about 2 months, depending on species

ORDER: carnivores

▽ *Lean and hardy, a dhole listens for danger. People sometimes call this animal the red dog of Asia.*

▽ *Stocky and short-legged, a bush dog of South America can scurry easily through dense underbrush.*

Dhole: 35 in (89 cm) long; tail, 17 in (43 cm)

Bush dog: 25 in (64 cm) long; tail, 5 in (13 cm)

litter a year. There are usually four to six furry, help-less pups in a litter. Both parents help care for the young. Some pups go off on their own before they are a year old. Others stay with their parents and help rear the next litter.

You can read about some of the better known wild dogs—coyotes, dingoes, foxes, jackals, rac-coon dogs, and wolves—under their own headings. Other wild dogs include the dhole (say DOLE) of Asia, the bush dog of South America, and the Cape hunting dog of Africa.

Dholes hunt in packs of five to twelve animals. In the forests of India, dholes run down axis deer and sambar deer. High in the mountains of China and Tibet, packs of dholes hunt wild goats and mountain sheep.

Bush dogs roam grassy swamps and tropical forests near rivers in South America. They probably live in family groups of three or four animals. Dur-ing the day, these bands hunt such rodents as pacas. The stocky bodies, short legs, and webbed paws of bush dogs are well suited to swimming. If their prey jumps into a river, bush dogs will dive in after it. Some dogs may already be waiting in the water for the prey.

Cape hunting dogs search for herds of ante-lopes and zebras on the plains and in the woodlands of eastern, central, and southern Africa. They usual-ly travel in packs of ten to twenty dogs. Before they hunt, the dogs jump around, make twittering calls, and lick each other's faces. It looks as if they are hav-ing a pep rally. Because they cooperate, Cape hunt-ing dogs are successful pack hunters.

All the adults in a pack of Cape hunting dogs help care for the pups. Pups beg for food from re-turning hunters. The adults then bring up meat they

◁ *Young wildebeest staggers as a pack of Cape hunting dogs closes in for the kill. When they hunt, these African wild dogs go after a slow animal. The pack chases the animal until it cannot go any farther.*

▽ *Pausing during its meal, a Cape hunting dog watches for hyenas that might try to steal its kill.*

Cape hunting dog: 40 in (102 cm) long; tail, 14 in (36 cm)

have swallowed and carried back in their stomachs. Old and lame animals and those that stayed behind to watch the pups are fed in the same way.

Domestic dogs are relatives of wild dogs. Scientists think all dogs may have descended from wolves or jackals. Thousands of years ago, people began to tame the wild dogs that came prowling around their camps, probably in search of food. Later some tame dogs may have become wild again. The dingo may be one kind of tame dog that became wild.

Today there are several hundred breeds of domestic dogs. They range in size from the tiny Chihuahua to the huge Irish Wolfhound. Their coats can be smooth and short or thick and shaggy. Some domestic dogs could not survive in the wild now. But domestic dogs and wild dogs still share many of the same traits.

Most dogs defend their territories fiercely. A

△ *In friendly play, Cape hunting dogs rear up and lunge at each other's throats. When excited, the dogs utter twittering, birdlike cries.*

Two Cocker Spaniels lounge on a tree ▷
stump. Popular family pets, these
animals were once used to hunt birds.

Old English Sheepdog: 22 in (56 cm) tall at the shoulder

Cocker Spaniel: 15 in (38 cm) tall at the shoulder

△ *Shaggy hair nearly hides the eyes of an Old English Sheepdog.*
Domestic dogs can vary greatly—from the tiny, smooth-haired
Chihuahua to the huge, hairy Irish Wolfhound. Despite the
differences among the several hundred breeds, all domestic dogs
belong to a single species, or kind, of dog.

▽ *Long legs of a Pharaoh Hound help it chase down animals. The*
rulers of Egypt may have used this kind of sleek racer to hunt
gazelles more than 3,000 years ago.

Pharaoh Hound: 25 in (64 cm) tall at the shoulder

Golden Retriever: 24 in (61 cm) tall at the shoulder

△ *Ears flying and eyes squinting, a*
Golden Retriever hits the water. A
retriever accompanies a hunter and
retrieves birds shot by its master. The dog
grasps the birds gently in its mouth.

pet dog's territory may be its owner's house and yard. A dog marks its territory by leaving urine on trees and rocks. By sniffing these scent posts, other dogs can tell that the territory is occupied. Many wild dogs bury meat from a kill and return to eat it later. Many pet dogs have the same habit. They bury bones. Domestic dogs often turn around several times before lying down. Some wild dogs do this, perhaps to trample down grass and form a bed.

Dogs communicate with each other by scent, by the positions of their bodies, by the expressions on their faces, and by the sounds they make. Dogs that live and hunt in groups have different ways of communicating than dogs that live and hunt alone.

Dogs growl, snarl, and whine. Most dogs bark as a warning. Domestic dogs may yelp in fear or in pain. Dogs howl to communicate over long distances. Bared teeth and bristling fur are clear threatening signs. A pup begs for food or for attention by rolling over, whining, and nuzzling. When greeting another pack member—or a member of a pet's human family—an excited dog wags its tail.

△ *Thick, matted coat covers a Komondor, a guard dog of Hungary. Many kinds of fierce, loyal dogs work as police dogs or watchdogs. Some kinds of domestic dogs serve as guide dogs for blind people.*

"Keep it moving!" From its perch on a fence, a sheep dog directs traffic at a New Zealand sheep farm. Sometimes it even climbs on the backs of sheep. Other dogs round up the animals on the open range.

Dolphin

Dolphin is another name for porpoise. Find out about porpoises on page 452.

Donkey

The donkey is a domestic ass. Read about asses on page 66.

Dormouse

(say DOOR-mouse)

Common dormouse: 3 in (8 cm) long; tail, 3 in (8 cm)

Snug in its nest and fast asleep, a common dormouse looks like a tiny ball of fur. As it hibernates through the winter, its body temperature falls. Its heart rate and breathing slow down. In spring, the animal will wake up and build another nest in thick shrubbery.
Because the common dormouse often lives in hazel bushes and fattens itself on the nuts, people sometimes call it the hazel mouse.

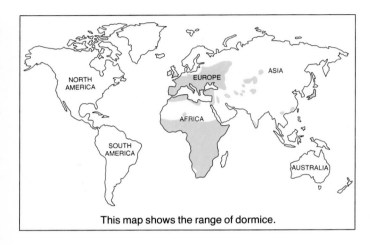

This map shows the range of dormice.

BEFORE ITS "LONG WINTER'S NAP," the dormouse fattens up on such food as nuts and fruit. Then, as cold weather sets in, this furry, squirrel-like rodent curls up in a burrow, a hole in a tree, or a space between rocks. There it hibernates (say HYE-burnates), or sleeps, for several months. The dormouse probably gets its name from a French word that means "to sleep."

In cold climates, dormice hibernate for nearly eight months—from late September to early May. Usually they hibernate alone. But sometimes several animals huddle together. From time to time, they wake up to eat seeds and nuts they have stored.

There are about twenty kinds of dormice. They live in Europe, Asia, and Africa. They are found in forests, grasslands, gardens, and parks. At night, the small animals scurry up and down bushes and trees, using their sharp claws to cling to branches. In the darkness, they search for insects, snails, and young birds.

Dormice also nibble at the fruit and nuts on trees. Farmers often wake to find their orchards damaged by these hungry animals.

Fat dormice, the largest of all dormice, are well known for their big appetites. In early summer, the

Garden dormouse: 5 in (13 cm) long; tail, 4 in (10 cm)

animals weigh between 3 and 4 ounces (85-113 g). Then they begin stuffing themselves with food. By winter, they have nearly doubled in weight.

As they hibernate, dormice use up their stored fat. When they wake up in the spring, they immediately set about searching for food.

Soon the females are ready to give birth. They make nests out of grass and leaves. After a three- or four-week pregnancy, they bear two to nine offspring. Although the newborn nurse for only three or four weeks, they often stay with their mothers during the following winter.

DORMOUSE

LENGTH OF HEAD AND BODY: 2-8 in (5-20 cm); tail, 2-6 in (5-15 cm)

WEIGHT: 1-4 oz (28-113 g)

HABITAT AND RANGE: bushes and brushy forests in Africa, Asia, and Europe

FOOD: fruit, nuts, insects, snails, young birds, and eggs

LIFE SPAN: 2 to 4 years in the wild

REPRODUCTION: 2 to 9 young after a pregnancy of 3 or 4 weeks

ORDER: rodents

◁ *Black mask around its eyes identifies a garden dormouse. Despite its name, this kind of dormouse sometimes nests in woods as well as in gardens.*

▽ *Stretching out its tail for balance, a fat dormouse scampers along a branch in search of figs. In summer, it eats so much that it nearly doubles its weight.*

Fat dormouse: 6 in (15 cm) long; tail, 5 in (13 cm)

Douroucouli

A douroucouli is a kind of monkey. Read about douroucoulis and other monkeys on page 376.

Dugong

(say DOO-gong)

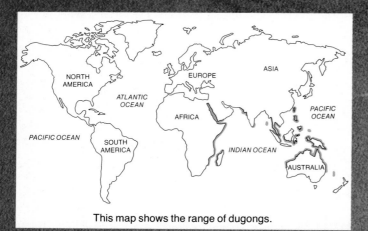

This map shows the range of dugongs.

DUGONG

LENGTH OF HEAD AND BODY: 8-10 ft (244-305 cm)

WEIGHT: 510-1,100 lb (231-499 kg)

HABITAT AND RANGE: warm, coastal waters scattered throughout the Indian Ocean and the western Pacific Ocean

FOOD: sea grasses

LIFE SPAN: 50 years in the wild

REPRODUCTION: usually 1 young after a pregnancy of about 1 year

ORDER: sirenians

Dugong calf shadows its mother as they swim in shallow waters off the coast of Australia. The calf's smooth, cream-colored skin will become rusty brown as it grows older. Dugongs usually bear one young after a year's pregnancy. The mother guards her calf closely during its first year and a half. She even carries it on her back from time to time.

186

PUSHED BY THE STRONG, WAVING MOTION of its tail and body, the torpedo-shaped dugong swims gracefully through the water. This huge animal — measuring about 9 feet (274 cm) long and weighing almost 600 pounds (272 kg) — lives in warm, coastal seas from Africa to Australia. Eating is its main activity, and sea grasses are its main food.

A dugong feeds alone or in a herd, day or night. It nibbles on underwater sea grasses. Using its flat, hairy snout, it roots for plants anchored on the bottom. It grasps the food with its coarsely bristled lips. Then it gives the plants a powerful shake that cleans off clinging grains of sand. Sometimes a dugong uses its flippers to scratch its face, to rub its gums, or to guide a young calf.

A dugong looks and acts much like its relative, the manatee. But its flat tail has a notch in the center, like a whale's tail. The manatee's tail is rounded. Dugongs and manatees also live in different parts of the world. Read about manatees on page 352.

Like all mammals, dugongs breathe air. They surface to take a breath every few minutes. But they can stay submerged for about six minutes.

For centuries, these marine mammals have been hunted for their skin, oil, and meat. Yet scientists do not know much about their habits. Today many countries have laws to protect dugongs.

Duiker

A duiker is a kind of antelope. Read about duikers and other antelopes on page 52.

E

Long-nosed echidna: 24 in (61 cm) long; tail, 4 in (10 cm)

Echidna

(*say* ih-KID-nuh)

Long-nosed echidna (above) searches the forest floor in New Guinea for soft soil in which to burrow. Tips of its sharp spines poke through thick hair. With a long hind claw (left), the short-nosed echidna can reach among its sharp spines and groom itself.

IS THERE A MAMMAL that lays eggs? Actually, there are two! One kind is the echidna, or spiny anteater, of Australia and New Guinea. The other is the platypus. Both these animals belong to the monotreme (say MON-uh-treem) order. You can read about the platypus on page 444.

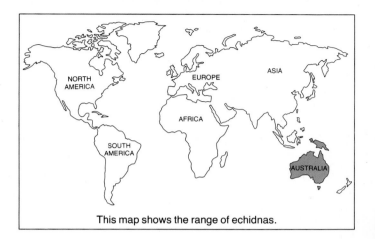

This map shows the range of echidnas.

ECHIDNA

LENGTH OF HEAD AND BODY: 14-39 in (36-99 cm); tail, 4 in (10 cm)

WEIGHT: 11-22 lb (5-10 kg)

HABITAT AND RANGE: forests, mountains, valleys, and plains of Australia and New Guinea

FOOD: earthworms, ants, termites, and other insects

LIFE SPAN: more than 50 years in captivity

REPRODUCTION: usually 1 young hatched from an egg after an incubation period of about 10 days

ORDER: monotremes

Short-nosed echidna: 16 in (41 cm) long; tail, 4 in (10 cm)

The female echidna carries a single leathery egg in a pouch that forms on her belly at the beginning of the breeding season. After about ten days, the egg hatches. The blind and hairless offspring, no bigger than a raisin, sucks milk from glands inside the pouch. The young grows quickly. After several weeks, sharp spines develop. Then it can no longer remain in its mother's pouch. After several years, the young is fully grown.

There are two kinds of echidnas—long-nosed and short-nosed. Both have long spines. They also have heavy claws and sensitive snouts. They use these to search for food. The short-nosed echidna looks for insects. It tears open logs and underground nests and can easily push over stones twice its weight. Its sticky tongue darts in and out, catching ants and termites.

The long-nosed echidna of New Guinea eats mostly earthworms. It hooks a worm on a row of tiny spines along a groove in its tongue. Then the echidna pulls the worm inside its beaklike snout.

When threatened, an echidna burrows straight down, sinking rapidly into the ground. Or it may squeeze into a hiding place. With only its spiny back showing, it is safe from danger.

Short-nosed echidna (above, right) digs a hiding place. After a few moments, only its prickly back will remain above ground. A young short-nosed echidna (right) waits in a sheltered spot for its mother. Spines have begun to appear on its back. Unlike all other mammals, except the platypus, newborn echidnas hatch from eggs. But an offspring still drinks its mother's milk.

Eland

The eland is a kind of antelope. Read about antelopes on page 52.

Elephant

(*say* EL-uh-funt)

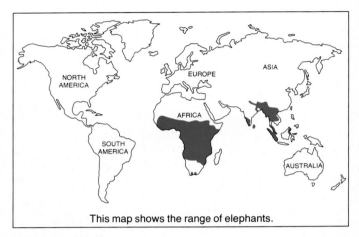

This map shows the range of elephants.

△ *Broken and worn, the tusks of an African elephant show the effects of years of heavy use. With their tusks, elephants dig for roots and move tree branches. They also use them to pry bark off tree trunks.*

Huge African elephant strides through the grass with a ▷ *cattle egret riding on its back. As they walk, elephants stir up insects, which the egrets eat.*

African elephant: 10 ft (305 cm) tall at the shoulder

LAZILY STRETCHING ITS TRUNK down to a stream, an elephant sucks up some water. It curls its trunk toward its mouth and squirts in the drink. Despite its usefulness in drinking, an elephant's trunk is more than a straw. The animal breathes through two nostrils at the end of its trunk. With the help of a fingerlike part at the tip, it can grasp small objects.

Elephants use their trunks mostly to drink and to bring food to their mouths. But they also can use them to nudge their calves or to pluck berries from a bush. If threatened, elephants may trumpet a warning through their trunks.

The largest of all living land animals, elephants may weigh more than 6 tons (5,443 kg). Thick skin crisscrossed with wrinkles covers their huge bodies. The animals have very little hair, just small clumps around their ear openings, on their chins, and at the ends of their tails.

Most elephants have tusks—huge, pointed, ivory teeth that grow all during an animal's life. Elephants use their tusks as tools. A tusk can help dig up a bush so the elephant can eat the roots. A tusk can pry bark from a tree. Often, one of an elephant's tusks is shorter than the other. The animal wears down that tusk by using it more than the other, just as people use one hand more than the other.

Millions of years ago, ancestors of the elephant roamed most of the earth. Scientists believe that Ice

Reaching high with its trunk, an African elephant ▷ pulls branches from an acacia tree. Elephants use their trunks to gather food and to get drinks of water.

Age people hunted these shaggy animals, called mammoths and mastodons. Drawings of these animals have been found in caves in Europe. Today wild elephants live only in Africa and in Asia.

The easiest way to tell the difference between an African elephant and an Asian elephant is by the ears. African elephants have much larger ears than their Asian relatives do. Their ears are shaped somewhat like the continent of Africa.

African elephants also are slightly taller and heavier. Male African elephants usually measure about 10 feet (305 cm) tall at the shoulder—2 feet (61 cm) taller than male Asian elephants. Both male and female African elephants have tusks that grow several feet long. Among Asian elephants, only the males grow tusks.

Although African and Asian elephants do not look exactly alike, their habits are similar. Both kinds of elephants feed mainly on roots, leaves, fruit, grasses, and sometimes bark. After pulling a bunch of grass from the ground with its trunk, an elephant

ELEPHANT

HEIGHT: 6-12 ft (183 cm-4 m) at the shoulder

WEIGHT: 5,000-14,000 lb (2,268-6,350 kg)

HABITAT AND RANGE: woodlands, grasslands, and forests in parts of Asia and Africa

FOOD: roots, leaves, fruit, grasses, and sometimes bark

LIFE SPAN: about 70 years in the wild

REPRODUCTION: usually 1 young after a pregnancy of 18 to 22 months

ORDER: proboscideans

In an elephant-style hug, two African elephants show affection by wrapping their trunks together.

Elephant

◁ *Eye-deep in water, an African elephant bathes in a river. The animal can stand underwater and breathe—as long as its trunk reaches above the surface.*

▽ *At a river in Kenya, African elephants cool off by taking a muddy bath. Elephants bathe often, because the hot sun dries their skin. During the hottest part of the day, they usually stay in the shade of trees. On very hot days, they use their huge ears like fans. As the temperature rises, they flap their ears faster.*

may beat the grass against its leg to shake the dirt off before eating. An adult elephant eats as much as 300 pounds (136 kg) of food a day!

To find that much food, these huge animals must roam wide areas. Females—called cows—travel together in herds with their young, called calves. Adult males—or bulls—usually travel alone or with other bulls. They join the group of females for mating and occasionally at other times.

An elephant cow usually gives birth to one calf

▽ *Stirring up a cloud of dust, African elephants powder themselves after bathing. First they suck dirt up with their trunks. Then they blow it over their wet bodies. The dirt helps protect their skin.*

△ *"Let's play!" a young African elephant seems to say to a resting friend. Young elephants often pull older elephants' tails or snatch food from their mouths.*

every two to four years. Elephants have the longest pregnancy of any mammal in the world—nearly 22 months. A newborn calf stands about 3 feet (91 cm) tall and weighs about 200 pounds (91 kg). It nurses for three or four years. About six months after birth, however, the calf begins to eat some solid food. Sometimes a calf sucks its trunk, just as a human baby sucks its thumb!

drink and bathe. Elephants travel to several water holes daily, and they sometimes spend hours rolling around in the mud and water. Surprisingly, elephants swim very well.

With their strong legs, elephants move around easily on land. They have been used as work animals for thousands of years. More than 2,000 years ago, they carried soldiers and weapons into battle.

Asian elephant: 8 ft (244 cm) tall at the shoulder

Small herd of Asian elephants travels with young in the middle and adults on either side.
Few Asian elephants remain in the wild. Governments have passed laws to protect them.

The youngster is looked after by other cows in the herd as well as by its mother. Young calves are often kept together in groups called kindergartens. One adult baby-sits while other adults feed.

Elephant calves play much of the time. They splash in the water, chase small animals, and fight each other with their trunks. When the calves stop playing they often lean against each other and nap.

Adult elephants need little sleep. They spend most of their time feeding or visiting water holes to

◁ *Drenching itself with its built-in hose, an Asian elephant takes an afternoon shower. Asian elephants are about 2 feet (61 cm) shorter than African elephants.*

In Asia, people still train elephants to lift and carry logs. Some elephants perform in circuses.

Because of their size, elephants have few enemies. Lions, tigers, crocodiles, and other meat-eating animals occasionally prey on small calves that have become separated from the herd. But these hunters rarely attack adult elephants. People are the elephant's only real enemies.

For centuries, people have killed elephants for sport and for their valuable ivory. In Asia, most wild elephants live on preserves. In Africa, some elephants still roam wild through grasslands and forests. Illegal hunting goes on today, even though laws have been passed in some places to prevent it.

Elk

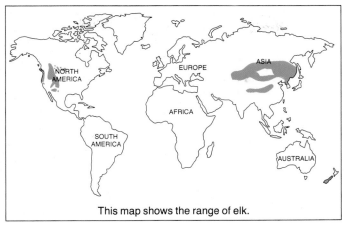

This map shows the range of elk.

IN WINTER, herds of elk feed in mountains and valleys throughout parts of western North America. Elk can paw through snow 2 feet (61 cm) deep to find grass, their favorite food. When the snow is deeper or encrusted with ice, elk must browse, or nibble at shrubs. The cold temperatures do not harm a healthy animal. Beneath an elk's heavy winter coat lies a layer of warm underfur. This woolly undercoat helps to hold in the animal's body heat.

Elk once roamed most of North America. But over the years hunters have killed off many of the animals. The spread of farming meant that elk had fewer places in which to live. Today the largest herds are found in Wyoming—on the National Elk Refuge and in Yellowstone National Park. Another kind of elk is found in central Asia.

Some people call elk *wapiti* (say woᴘ-ut-ee), an American Indian word meaning "light-colored deer." Elk are members of the deer family. Find out about deer on page 170.

Elk are larger than most other kinds of deer. Full-grown male elk, called bulls, weigh as much as 1,100 pounds (499 kg) and measure 5 feet (152 cm) tall at the shoulder. Female elk, called cows, weigh less and are shorter than bulls.

Only bulls have antlers. These large, heavy, bony growths develop from the top of an elk's head. They may measure 4 feet (122 cm) from base to tip.

In a land blanketed by fresh snow, an elk feeds near hot springs in Yellowstone National Park. The bull paws at the soft snow to clear patches of grass.

199

Every March, bulls shed their antlers. In May, new sets begin to grow. Rounded bumps start to swell on the bulls' heads, pushing up about half an inch (13 mm) each day. The antlers are covered with soft skin called velvet. New antlers develop each year.

As the weather becomes warm, elk begin to lose their long winter coats. Sleek, reddish coats replace their heavy gray-brown ones.

Soon the elk migrate, or travel, from the lower slopes to high mountain meadows. Some travel as far as 40 miles (64 km). Cows and their offspring move together. Bulls travel separately.

At their summer feeding grounds, cows give birth to young. A cow usually bears a single calf. The wobbly, spotted calf can stand up when it is about twenty minutes old. Within an hour, it begins to nurse. When the calf is several weeks old, it follows its mother and joins the herd of cows.

Late summer signals the mating season. Restless bulls thrash at bushes with their full-grown antlers to remove the velvet. As they search for mates, bulls make bugling sounds. They start on low notes that become high pitched and end in grunts. Sometimes bulls parade side by side. Often they fight over

Resting in the grass, a bull chews his cud. Soft skin called velvet covers his branching antlers.

Male elk swims behind two cows and their calves at ▷ mating time. If rivals come near, he may fight them with his sharp-tipped antlers, now bare of velvet.

the cows. Pushing and shoving with their antlers, they begin to wrestle. Their powerful neck muscles bulge. The battles usually do not end in injury. The weaker bull trots away.

The winning bull becomes master of a group of cows. He herds the cows together and fights off rival bulls. He has little time to eat or sleep. After a few weeks, he becomes exhausted, and another bull may take over the group.

The mating season lasts from late August into November. Then the leaders of the herds give up control of their cows, and all the elk gather once more into large groups. As the snow begins to fall, the elk again head for their pastures in the valleys.

ELK

HEIGHT: 4-5 ft (122-152 cm) at the shoulder

WEIGHT: 325-1,100 lb (147-499 kg)

HABITAT AND RANGE: mountain forests and grassy valleys of western North America and central Asia

FOOD: grasses, herbs, shrubs, and trees

LIFE SPAN: 8 to 12 years in the wild

REPRODUCTION: usually 1 young after a pregnancy of 8 or 9 months

ORDER: artiodactyls

▽ *Four-month-old calf nuzzles against its mother in a mountain meadow. The female grooms her calf's coat.*

Ermine The ermine is a kind of weasel. Read about weasels on page 564.

F

Ferret

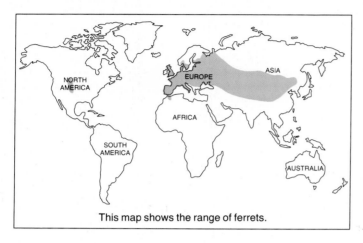

This map shows the range of ferrets.

FERRET

LENGTH OF HEAD AND BODY: 15-18 in (38-46 cm); tail, 5-7 in (13-18 cm)

WEIGHT: about 3 lb (1 kg)

HABITAT AND RANGE: prairies, forests, and meadows of Europe, Asia, North Africa, and the western United States

FOOD: prairie dogs, ground squirrels, and other rodents, as well as amphibians, reptiles, birds, and fish

LIFE SPAN: 8 to 14 years in captivity, depending on species

REPRODUCTION: 2 to 12 young after a pregnancy of 6 weeks, depending on species

ORDER: carnivores

RAREST OF ALL NORTH AMERICAN MAMMALS, the black-footed ferret has nearly disappeared from the plains of the central United States.

With a black band across its eyes, the sleek black-footed ferret looks like a masked bandit. Its fur

Sniffing with its sensitive nose, a domestic ferret ▷ prepares to leave the nest it shares with its mate. People in Europe once used ferrets to hunt other small animals. Now they sometimes keep ferrets as pets.

is tan, but black hair covers its feet and legs and the tip of its tail. Shaped like its relative the weasel, the ferret has a long, slender body. When it hunts prairie dogs, it easily slides into a burrow.

Though the ferret also eats mice, rabbits, and ground squirrels, prairie dogs are its main food. Ferrets usually live in prairie dog towns. These are large areas where prairie dogs dig networks of burrows in which they live and raise young. There, the black-footed ferret can find and kill its prey. It also finds shelter in abandoned burrows.

In the last few decades, the number of prairie dogs has decreased. Some farmers and ranchers have tried to get rid of prairie dogs because they burrow holes in fields and pastures. Since there are fewer prairie dogs, black-footed ferrets may be dying out.

Scientists think that a few ferrets exist in some remaining prairie dog towns. The surest sign of a ferret is a long trench extending from a burrow entrance. During the day, while a ferret sleeps underground, a prairie dog may close up the entrance to the burrow in which its enemy is resting. It kicks dirt into the opening and pushes it down with its nose. At night, the ferret can dig itself out, often leaving the telltale trench.

Black-footed ferret seeks—and finds—its prey in a prairie dog town in South Dakota. Slinking up to a burrow, the ferret sniffs carefully. Smelling a prairie dog, it enters the burrow and searches the tunnels.

Relatives of the black-footed ferret live in Europe and in Asia. The black-footed ferret's closest relative is the steppe polecat. It makes its home on the plains of central Asia and of eastern Europe. The steppe polecat hunts at night. It eats mainly rodents, especially ground squirrels. If it kills more than it can eat, it stores the extra food in its burrow.

Another relative, the European polecat, is found in Europe. It often lives near farms, hunting mice and rats at night. It has dark fur with a light-colored undercoat.

Some people raise domestic, or tame, ferrets for their soft fur—called fitch. It is used in making coats. For centuries, people in Europe also used ferrets to hunt small animals. The ferret was sent down a burrow to chase out the inhabitant. From this practice comes the expression "to ferret out," which means to uncover something hidden.

After a successful hunt, the ferret pops its head out, holding a prairie dog firmly by the throat. Then it crawls out of the burrow and carries off its prize across the prairie.

Black-footed ferret: 16 in (41 cm) long; tail, 6 in (15 cm)

Fisher
The fisher is a close relative of the marten. Read about both animals on page 363.

Flying lemur

(*say* FLY-ing LEE-mur)

Malayan flying lemur: 15 in (38 cm) long; tail, 9 in (23 cm)

Cradled in the folds of its mother's silky fur, a young flying lemur stays warm and dry. When its mother glides to another tree, the youngster clings to her belly.

FLYING LEMUR

LENGTH OF HEAD AND BODY: 13-17 in (33-43 cm); tail, 7-11 in (18-28 cm)

WEIGHT: 2-4 lb (907-1,814 g)

HABITAT AND RANGE: forested areas and coconut plantations in parts of Southeast Asia

FOOD: leaves, buds, and flowers

LIFE SPAN: unknown

REPRODUCTION: usually 1 young after a pregnancy of about 60 days

ORDER: dermopterans

HIDDEN FROM VIEW and sheltered by a roof of leaves, the flying lemur spends most of its life high in the trees. The flying lemur is rare and is hard to find in its habitat—forests and coconut plantations in parts of Southeast Asia. And since the flying lemur does not live long in captivity, little is known about the animal's way of life. Another name for the flying lemur is colugo (say kuh-LOO-go). The flying lemur is not related to the lemur, which is a member of the primate order.

During daylight hours, most flying lemurs rest. Some of them hang from branches and from huge palm leaves, gripping firmly with their sharp, curved claws. Others curl up in holes in tree trunks. Their brownish gray fur—speckled with white—blends in with the bark of the trees, so the animals are hard to see.

At dusk, flying lemurs begin to eat. Trees are their main source of food. The animals feed on leaves, buds, and flowers. They probably get water by licking rain from the leaves.

A flying lemur travels from tree to tree by gliding. Spreading its limbs, the flying lemur stretches folds of skin that extend from the sides of its neck to all four feet and its tail. Then it leaps into the air and glides down to another tree. From below, the animal looks like a kite. Some flying lemurs may cover as much as 200 feet (61 m) in a single leap!

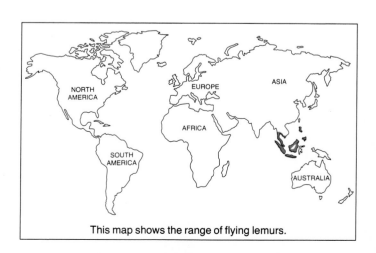

This map shows the range of flying lemurs.

Fossa

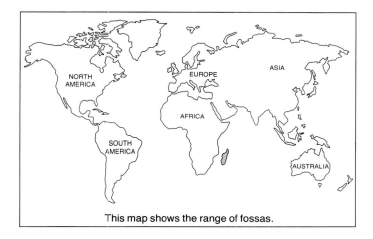

This map shows the range of fossas.

FOSSA

LENGTH OF HEAD AND BODY: 24-31 in (61-79 cm); tail, 24-31 in (61-79 cm)

WEIGHT: 15-26 lb (7-12 kg)

HABITAT AND RANGE: forests of Madagascar

FOOD: small mammals and birds

LIFE SPAN: 15 years in captivity

REPRODUCTION: 2 to 4 young after a pregnancy of about 3 months

ORDER: carnivores

CREEPING NOISELESSLY along a branch, a fossa stalks its prey. It hunts at night, catching birds and lemurs by surprise.

The fossa is the largest predator, or meat-eating hunter, on Madagascar, an island off the southeastern coast of Africa. This sleek animal is about twice as large as a house cat. With sharp teeth and curved claws, the muscular fossa can overpower almost any animal it attacks.

Whether it is looking for food or resting in the fork of a tree, the fossa spends much of its time high above ground. Its claws and the hairless pads on its feet give the fossa a secure grip in the trees. Its long, heavy tail helps the animal keep its balance.

Like their close relatives the civets, fossas communicate with scent. A fossa has glands that produce a strong-smelling oil. As it moves about, the fossa occasionally leaves some of this oil along its trail. During September and October, the oily patches serve as scent signals as fossas look for mates. During most of the year, however, fossas avoid each other. Read about civets on page 154.

Three months after mating, a female fossa finds a hollow tree or a small cave in which to give birth. Her litter may contain two to four offspring. The young are full grown in about four years.

Catlike fossa rests on a tree limb. This strong meat eater hunts on the ground and in the trees.

Fox

THE THICK, BUSHY TAIL of a fox has many uses. When a fox curls up in cold weather, it curls its tail— sometimes called a brush—to cover its feet and its nose. The tail looks like a long, woolly scarf. When a fox runs, its tail streams out behind. This helps the fox keep its balance as it zigzags across the land. By moving its brush in certain ways, a fox can send messages to other foxes. But it is not true that a fox uses its brush to sweep away its tracks in snow.

Foxes live nearly everywhere in the world. The arctic fox stays in the treeless regions of the Far North. Another kind of fox, the fennec (say FEN-ick), makes its home in the deserts of Africa and in the

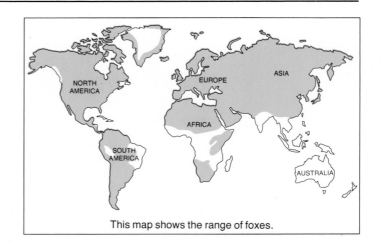

This map shows the range of foxes.

FOX

LENGTH OF HEAD AND BODY: 14-39 in (36-99 cm); tail, 7-20 in (18-51 cm)

WEIGHT: 3-29 lb (1-13 kg)

HABITAT AND RANGE: almost every kind of habitat worldwide, except for Antarctica

FOOD: small mammals, birds, eggs, insects, reptiles, amphibians, fish, grasses, berries, nuts, roots, and the remains of dead animals

LIFE SPAN: as long as 13 years in captivity, depending on species

REPRODUCTION: 2 to 12 young after a pregnancy of 1½ to 2½ months, depending on species

ORDER: carnivores

◁ *Extending its tail, a red fox in Maine follows the track of its prey. Foxes hunt mostly at night.*

Red fox: 31 in (79 cm) long; tail, 15 in (38 cm)

△ *Female red fox nuzzles her two-month-old pup. Most adult foxes care for their young for about six months. Then the offspring go off to live alone. Newborn red foxes have soft brown or gray fur. Their colorful red coats do not begin to grow in until they are about four weeks old.*

◁ *Up on a stump, a red fox sniffs for traces of other foxes. Foxes make scent posts by marking trees, rocks, or patches of ground with their urine. From the smell, a fox knows that another fox has passed by.*

207

◁ Swift fox pauses on a prairie in Colorado. This animal lives up to its name. It can escape from a coyote by sprinting to a burrow. It often catches jackrabbits.

▽ Enormous ears make a tiny fennec look even smaller. The ears help the animal hear prey as well as approaching enemies. Thick fur insulates this fox from the desert's nighttime cold.

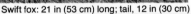

Swift fox: 21 in (53 cm) long; tail, 12 in (30 cm)

Fennec: 15 in (38 cm) long; tail, 10 in (25 cm)

Middle East. The chilla (say CHILL-uh) roams dry plains in South America. The red fox lives in Europe, Africa, and Asia, as well as in North America.

Not all red foxes are red. They vary in color from pale rosy gold to deep rusty brown. Some of the animals have pure black fur. When their fur is black and frosted with white, red foxes may be called silver foxes. Cross foxes have red coats with black crosses on their backs and shoulders. Hunters sometimes trap foxes for their soft, thick fur. Foxes are also raised on fur farms. Silver fox fur is especially prized, as is the fur of the arctic fox.

Arctic foxes can be white or blue. Both kinds change color with the season. Like the land around them, white foxes turn brown in summer. They

◁ Chilla stretches for a mouthful of berries. This yellow-gray fox roams the dry plains of South America.

Chilla: 31 in (79 cm) long; tail, 13 in (33 cm)

◁ *Rare Simien fox of Ethiopia stands alert, searching for prey. It feeds mainly on rodents. This fox lives on plateaus and in mountains as high as 10,000 feet (3,048 m). Although these foxes hunt alone, they live in groups and may play and romp together.*

Simien fox: 31 in (79 cm) long; tail, 13 in (33 cm)

▽ *Litter of bat-eared foxes in Africa basks in the sunshine. Their mother will use a high-pitched whistle to call them. Bat-eared foxes feed mainly on insects, especially termites, which they dig from the ground.*

Bat-eared fox: 22 in (56 cm) long; tail, 12 in (30 cm)

Gray fox explores a hollow log. It searches for insects ▷ *to eat or for a place to make a den. Gray foxes roam dense, brushy forests in North and South America.*

Gray fox: 28 in (71 cm) long; tail, 13 in (33 cm)

April sun bathes an arctic fox resting on a snowy hillside in Alaska. Prey can be hard to find during the long, northern winter. The foxes often follow polar bears onto pack ice and eat scraps left from the bears' kills. In spring, the foxes catch rodents, birds, and fish.

Fox

blend in so well with their surroundings that it is hard to spot the animals against the golden grass and bare ground. Blue foxes change from a light blue-gray coat to a darker one in summer.

Other foxes match their backgrounds, too. The sand-colored kit fox fades into the desert terrain of the western United States. The tan swift fox matches the color of dry prairie grasses. The gray fox almost disappears among the rocks and leaves of shadowy forests in North and South America. Foxes are adapted, or suited, to their surroundings in other ways. Long fur on the bottoms of the arctic fox's broad paws protects its feet from the cold. Sharp claws help the arctic fox keep its footing on icy ground. The fennec and other desert foxes also have furry feet to help them run in soft sand.

The ears of the fennec and of the bat-eared fox are very different from those of the arctic fox: They are enormous, and they help the desert animals hear prey burrowing in the sand. Because their ears are so big, these foxes lose body heat through them. This helps them keep cool. In contrast, the small, round ears of the arctic fox barely poke above its long, thick coat. Because its ears are only slightly exposed to the cold, the animal loses very little heat through them.

Red foxes often live where forests and farmlands meet. Woods offer good sites for fox dens. In fields, foxes can find rabbits, mice, and birds for food. Red foxes also eat lizards, frogs, and fish. If farms are nearby, the foxes may raid hen houses for chickens and eggs. In summer, they feast on blackberries and plums. They gobble down grapes and nuts in the fall.

Most foxes hunt alone and live together only when they are raising young. But some foxes live in pairs or in groups all year round. Although some foxes dig their own dens, others may take over abandoned burrows. Or they may creep under thick bushes or into hollow logs. With the help of sharp, curved claws, gray foxes can climb straight up a tree until they find a hollow place in which to hide. Desert foxes dig networks of tunnels under the sand. Their dens remain cool even in the hottest weather. During a blizzard, arctic foxes may dig into the snow for shelter.

Once a year, a female fox bears two to twelve helpless pups. The mother nurses the pups for about two months. The father also helps to care for them. As long as the pups are too young to be left alone, he carries prey to the den for the mother. Later both parents go hunting to feed their pups.

Fox pups are playful. They race and tussle. They bark and yap. They crouch, then jump on one another—or on anything that wiggles, hops, or flies. The parents bring home live rodents for the pups to chase and to pounce on. This gives them practice in hunting.

Soon the pups tag along as their parents search for food. At about six months of age, young foxes leave to seek hunting grounds of their own.

A fox hunts different animals in different ways. It catches mice by pouncing when it hears rustling sounds in the grass. A kit fox may stand on its hind legs and turn in a circle, ready to leap the instant a mouse gives itself away. Bat-eared foxes dig for insects in the sand. To catch a bird, a red fox creeps up silently until it is close enough to spring. It can run down a rabbit in a wild, zigzag chase.

To outwit its own enemies, a red fox doubles back on its tracks. Bat-eared foxes listen for danger, then scurry under a rock or into their sandy burrows in the desert. An arctic fox always keeps a sharp lookout for wolves as it feeds.

▽ *Poking its head out of its den, an arctic fox pup in Alaska shows its brown summer coat. The animal's fur will become thick and white before winter. An arctic fox digs out a new chamber in its den once a year for each new litter of pups.*

Arctic fox: 24 in (61 cm) long; tail, 12 in (30 cm)

G

Galago

Galagos are bush babies. Read about them on page 112.

Gazelle

(say guh-ZELL)

Thomson's gazelle: 25 in (64 cm) tall at the shoulder

△ *Less than a day old, a gazelle nuzzles against its mother to nurse. The female Thomson's gazelle licks her offspring to clean it.*

Crowned with slightly curved, ringed horns, a male ▷ *Thomson's gazelle chews a blade of grass on an African plain. Of all the gazelles, only the Thomson's gazelle has jet-black stripes on its sides.*

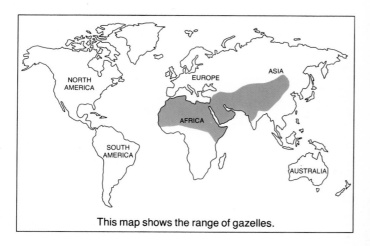

NORTH AMERICA

EUROPE

ASIA

AFRICA

SOUTH AMERICA

AUSTRALIA

This map shows the range of gazelles.

ALERT TO DANGER, a herd of Thomson's gazelles grazes on an open plain. The animals look up from time to time while feeding. Suddenly, a few members of the herd sight enemies—a pack of wild dogs. The gazelles jump into the air. Bouncing up and down with stiff, straight legs, they signal an alarm to the rest of the herd. Soon the whole herd is bouncing in this way. Then the gazelles run off, scattering in all directions. One gazelle may be caught by the dogs, but most of the herd will escape. To get away from a cheetah, the fastest land mammal, gazelles usually gallop away without bouncing. By running at speeds of about 40 miles (64 km) an hour, they can sometimes avoid lions, leopards, hyenas, wild dogs, and other meat eaters.

Thomson's gazelles are one of about 14 species,

or kinds, of gazelles—a group of medium-size antelopes. Gazelles are found in Asia and in Africa. Some live in mountainous regions. Others make their homes in the desert or in open shrubby areas called the bush. Most gazelles live on grassy plains.

Gazelles grow ringed horns that may be straight or curved. In many species, both males and females have horns. A female's horns are shorter and thinner than those of a male. Their coats range from grayish white to orange to brown. Some gazelles have a dark band that runs along each side of their bodies. Bellies and rumps are usually white.

Gazelles roam in herds numbering from a few animals to several hundred. They graze on grasses or nibble on leaves and shoots of bushes and other

◁ *Watchful Grant's gazelles stare across dry grass in Kenya. These animals need little moisture to survive. In the dry season, they may remain on the plains long after other animals have left in search of water.*

plants. During the rainy season, when food is plentiful, herds of gazelles may come together to form groups of thousands of animals. In the dry season, some gazelles migrate, or travel, from the plains to the bush in search of food and water. Certain other gazelles that are able to survive without drinking water remain behind.

At certain times of the year, an adult male of some species leaves the herd and marks a territory,

GAZELLE

HEIGHT: 20-43 in (51-109 cm) tall at the shoulder

WEIGHT: 26-165 lb (12-75 kg)

HABITAT AND RANGE: grasslands, treeless plains, shrubby areas, deserts, and mountainous regions of parts of Africa and Asia

FOOD: grasses, herbs, leaves, buds, and shoots

LIFE SPAN: as long as 17 years in captivity, depending on species

REPRODUCTION: 1 or 2 young after a pregnancy of about 5 or 6 months, depending on species

ORDER: artiodactyls

Grant's gazelle: 35 in (89 cm) tall at the shoulder

◁ *With heads lowered, two male Grant's gazelles prepare to test their strength. The animals may fight often, but they rarely hurt each other.*
▽ *Dorcas gazelles race across a rocky hill. These rare animals live in desert areas of northern Africa.*

Dorcas gazelle: 24 in (61 cm) tall at the shoulder

215

or area, of his own. He may fight to keep rival males out of his area. He tries to mate with the females that wander into his territory. Females and young stay together in groups. Males without territories roam in bachelor herds.

About six months after mating, a female leaves the herd and gives birth to one or two young. For sev-eral days or even weeks, the young gazelle lies hid-den in the grass. Its mother goes off to feed as usual, returning several times a day to nurse her offspring. When the young gazelle is old enough, it begins to follow its mother and to graze with the herd. If the offspring is a female, she will stay with a herd of fe-males. Male offspring will join a bachelor herd.

Gemsbok

The gemsbok is a kind of antelope. Find out about antelopes on page 52.

Genet

(say JEN-ut)

SLEEK AND SLENDER, the genet glides through thickets and between rocks on dry grassy plains. In dense forests, the animal darts up trees and leaps nimbly among the branches. Several species, or kinds, of genets live in Africa. One kind is also found in the Middle East and in southern Europe.

GENET

LENGTH OF HEAD AND BODY: 16-23 in (41-58 cm); tail, 13-21 in (33-53 cm)

WEIGHT: 2-7 lb (1-3 kg)

HABITAT AND RANGE: forests and grasslands of Africa and parts of southern Europe and the Middle East

FOOD: small animals and fruit

LIFE SPAN: less than 10 years in the wild

REPRODUCTION: 2 to 4 young after a pregnancy of 2½ months

ORDER: carnivores

This map shows the range of genets.

People know little about these animals because they rarely see them in the wild. Genets sleep dur-ing the day. They curl up in hollow trees, under bushes, or in tall grass. Their tan fur with light-and-dark markings blends into their surroundings.

After dark, genets prowl in search of food. Slinking through the forest, they eat insects, fruit, and birds. On the plains, they catch lizards and snakes. They also feed on mice and rats. People in ancient Egypt kept genets in their homes just to catch these small rodent pests.

Except for their pointed noses and short legs, genets look like small cats. Actually they are a kind of civet. Find out more about other civets on page 154.

The genet uses scent to communicate with oth-er genets. It has a scent gland under its long, ringed tail. It marks its path with sweet-smelling oil from this gland. By the odor one genet leaves, other gen-ets know it has passed by.

In captivity, female genets can have two litters of two to four offspring each year. The young nurse for a couple of months. When they are six weeks old, they begin to eat solid food. In the wild, genets prob-ably live less than ten years. In zoos, they may sur-vive twice as long.

Bold markings on a common genet's body and tail stand out against the darkness. Although experts believe that there are many genets in Africa, people rarely see these silent, shy animals. Sometimes only the gleam of its large eyes, caught in the beam of a flashlight, gives the genet away.

Common genet: 20 in (51 cm) long; tail, 18 in (46 cm)

Gerbil

Gerbil's tufted tail sweeps across desert sand in Africa. The animal's coloring blends with its surroundings.

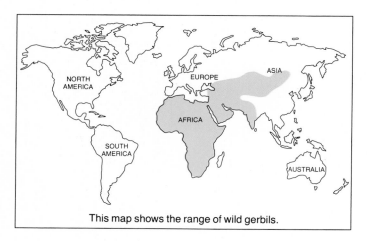

This map shows the range of wild gerbils.

NOT LONG AGO, most Americans had never heard of the small, frisky gerbil. In the 1950s, a scientist in New York got several Mongolian gerbils for his work. Today most of the pet gerbils in the United States trace their ancestors back to his nine animals!

Mongolian gerbils are actually jirds, close relatives of about a hundred other kinds of wild gerbils.

Most of these rat-size animals are not kept as pets. They usually live in dry parts of Asia and Africa. Their gray or sand-colored fur blends well into their environment. Gerbils often hop across the sand at night, using their long, thin tails for balance. Or they scurry about, looking for seeds, leaves, roots, flowers, and insects. Their food supplies them with almost all of the water they need.

Gerbils look for hard foods to gnaw on. This keeps their teeth from getting too long. Like other rodents' teeth, their front teeth never stop growing.

During the day, most gerbils escape the heat by staying in their underground homes. Some live in small, simple burrows. Others dig complex tunnel systems. Often several gerbils form a colony by building their burrows close together.

In nests inside their burrows, females give birth to litters of one to eight young. Because a pregnancy lasts less than a month, gerbils are able to have several litters a year. The newborn stay underground for about three weeks after birth. Then they begin to search for food on their own.

Round entrance (above) leads to the tunnels of a gerbil burrow in Africa. Underground, two sleeping gerbils nestle (right), away from the daytime heat. At night, they will look for insects, seeds, and other plant foods. They will carry some of their finds back to their burrows and store them there to eat later.

GERBIL

LENGTH OF HEAD AND BODY: **3-7 in (8-18 cm); tail, 3-9 in (8-23 cm)**

WEIGHT: **1-7 oz (28-198 g)**

HABITAT AND RANGE: **mostly dry, sandy areas of Asia and Africa; some kinds of gerbils are kept as pets**

FOOD: **seeds, roots, leaves, flowers, and insects**

LIFE SPAN: **about 4 years in the wild**

REPRODUCTION: **1 to 8 young after a pregnancy of about 3 weeks**

ORDER: **rodents**

Gerenuk

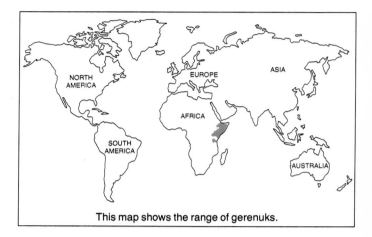

This map shows the range of gerenuks.

GERENUK

HEIGHT: 35-41 in (89-104 cm) at the shoulder; about 7 ft (213 cm) when standing

WEIGHT: 75-110 lb (34-50 kg)

HABITAT AND RANGE: dry brushy areas of eastern Africa

FOOD: leaves, shoots, flowers, and fruit of woody plants

LIFE SPAN: as long as 11 years in captivity

REPRODUCTION: 1 young after a pregnancy of about 7 months

ORDER: artiodactyls

WANDERING SLOWLY among thornbushes, a gerenuk searches for food. The animal rises gracefully on its hind legs. Standing straight and tall, it rests its forelegs against a bush. It chooses a leaf and plucks it with its flexible lips. By stretching its long neck, it can feed on the leaves of high branches that smaller animals cannot reach.

Gerenuks live in dry brushy areas of eastern Africa. During the day, they browse, that is, they nibble on leaves and shoots of woody plants. Occasionally, they feed late at night.

Gerenuks seem to get most of the moisture they need from the plants they eat. Some scientists think that gerenuks never drink water at all!

A male gerenuk weighs about 100 pounds (45 kg). It measures more than 3 feet (91 cm) tall at the shoulder. But when standing upon its hind legs, the animal reaches a height of about 7 feet (213 cm). Its curved, ringed horns grow to a length of about 14 inches (36 cm). A female gerenuk is smaller and has no horns.

Gerenuks are closely related to gazelles. The

△ Up on its hind legs, a male gerenuk nibbles leaves from a thorny shrub. For support, the animal leans its front legs lightly against the branches.

Like living statues, female gerenuks in Kenya freeze ▷ and watch for danger. With their long, slender necks, they can see over tall bushes and shrubs. Their name in an African language means "giraffe-necked."

gerenuk's long neck, however, reminds people of a giraffe. So the gerenuk is sometimes called a giraffe gazelle. Read more about gazelles on page 212.

Gerenuks have keen senses. Looking and listening, they stay alert to enemies such as lions and leopards. If startled, gerenuks freeze. They stand perfectly still for several minutes. Their thin legs are difficult to see among trees and bushes. Their reddish brown coloring sometimes blends with their surroundings. If an enemy approaches, the animals dash away. They may try to escape their pursuer by dodging around trees and bushes.

An adult male gerenuk holds a territory, or area, a few square miles in size. The male marks bushes and trees in this area with a substance from glands in the corners of his eyes. A small group of females and their young usually lives and feeds in this territory. Sometimes the male stays apart from them. Occasionally females also wander on their own.

A male usually mates with the females in his territory. Gerenuks may mate at any time of year. About seven months later, a female gerenuk gives birth to one young. The newborn stays hidden in a sheltered place for a month or more. During the day, its mother goes off to feed. When she returns, she nurses her offspring.

If the young gerenuk is a female, she might remain with her mother. But if the offspring is a male, he will leave his mother after about a year. Then he may roam with a few other young males. When he is fully grown, he will try to find a territory of his own.

Gibbon: 24 in (61 cm) long

Long-armed gibbon hangs lightly between two small branches. These apes spend most of their time in trees.

GIBBON

LENGTH OF HEAD AND BODY: 16-36 in (41-91 cm)

WEIGHT: 9-29 lb (4-13 kg)

HABITAT AND RANGE: tropical forests of southern and southeastern Asia

FOOD: fruit, flowers, leaves, insects, birds' eggs, and young birds

LIFE SPAN: about 34 years in captivity

REPRODUCTION: usually 1 young after a pregnancy of about 7 months

ORDER: primates

Gibbon (*say* GIB-un)

This map shows the range of gibbons.

GRACEFUL, ACROBATIC APES, gibbons live high in the trees of tropical forests in southern and southeastern Asia. Like other apes—gorillas, orangutans, and chimpanzees—gibbons have no tails. But these hairy animals are smaller than their relatives. So gibbons are called lesser apes. A full-grown male gibbon measures less than 3 feet (91 cm) long. It weighs about 15 pounds (7 kg). Female gibbons are slightly smaller and lighter.

A gibbon is well equipped for life in the trees. Its bones do not weigh much. And its long, strong arms are twice the length of its body. The animal has very long fingers and shorter thumbs. When the gibbon swings through the branches, the thumb is out of the way. A gibbon uses its fingers almost like hooks.

Gibbons travel through the trees with a hand-over-hand movement called brachiation (say bray-kee-AY-shun). Most of the time, gibbons move along at a leisurely pace. But when necessary, they can swing at an astonishing speed.

Sometimes gibbons walk upright along branches. To keep their balance, they stretch their arms out to the sides. They look a little like tightrope walkers. They can cross large gaps between branches—sometimes covering 30 feet (9 m) in a single leap! Occasionally, gibbons come down to the ground.

Gibbons usually eat fruit, flowers, and leaves.

White patches of dense, woolly hair highlight the face, ▷ hands, and feet of a gibbon. Strong flexible fingers give the animal a hooklike grip.

Gibbon

Throat sac inflated like a balloon, a siamang roars a warning to other gibbons: "Stay out of my territory!" A siamang's booming cry can be heard more than a mile away.

than other kinds of gibbons. A male weighs about 22 pounds (10 kg). The siamang has a special throat sac to make its voice carry over long distances. The animal puffs up the sac until it is about the same size as its head. Then it makes an amazingly loud "hoo-hoo-hoo" sound. Its calls can be heard more than a mile away.

Within their own territories, gibbons have favorite trees to sleep in. They return to the same trees night after night. There they huddle together. Or they may sleep alone, wedged in a tree fork. Unlike other apes, gibbons do not build sleeping nests. Special pads of tough skin on their hind ends serve as built-in cushions.

◁ *Young siamang imitates its parent by dangling between tree branches. Adult siamangs are larger and darker than other kinds of gibbons. They have dense black hair.*
▽ *Hanging carefully just above the water's surface, a siamang reaches to get a drink. The animal dips in its hand, then licks the water from its fingers.*

But they also eat insects, birds' eggs, and probably young birds. They live in family groups that include a male, a female, and two to four offspring.

A gibbon usually has only one mate during its lifetime. About seven months after mating, females give birth to a single young. The newborn travels with its mother soon after birth, clinging to the hair on her belly. Young gibbons stay with their parents about six years. They go off by themselves when they are ready to start families of their own.

Each gibbon family occupies a well-defined area in the forest. The adults defend this territory against other gibbons. Every morning when the gibbons wake up, they sing loudly to remind other animals that this is their area. Other gibbons should stay away! If they do come near, the family challenges the intruders by hooting loudly.

The kind of gibbon that makes the loudest noise is the siamang (say SEE-uh-mang). It is also larger

Siamang: 30 in (76 cm) long

Giraffe

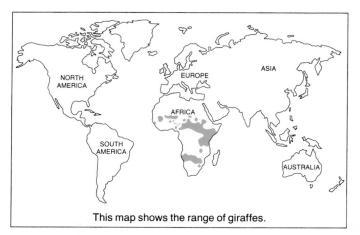

This map shows the range of giraffes.

IT'S IMPOSSIBLE TO CONFUSE the long-legged, long-necked giraffe with any other mammal. Giraffes are the tallest animals on earth. Adult males, called bulls, grow to a height of nearly 19 feet (6 m). Their legs are about 6 feet (183 cm) long. And their necks can be even longer than that!

Giraffes live on the savannas, or grasslands, of Africa. The animals wander in small groups of about five members. Giraffes live together peacefully most of the time. But sometimes bulls have contests to see which animal is stronger. During these "necking" bouts, bulls slam their necks and heads together until one gives up.

Stretching its long neck, a giraffe can eat leaves from the tops of thorny acacia trees. The animal pulls the leaves from a branch with its tongue, which can measure 21 inches (53 cm) long. Because of stinging ants among the branches, the giraffe does not linger at a tree.

Giraffes can go for several days without water. They get moisture from the juicy leaves of the plants they eat. When a giraffe does try to drink from a stream or a water hole, it usually spreads its front legs far out to the sides. Then it stretches its neck down as far as it can and takes a drink. In this awkward position, it is hard for a giraffe to watch for danger.

Towering above acacia trees, three giraffes move to a new feeding area on the grasslands of Africa. Giraffes often travel many miles every day in search of food. Each animal eats hundreds of pounds of leaves each week.

Giraffe: 17 ft (5 m) tall

Giraffe

Even though a giraffe's neck is very long, it has only seven neck bones—the same number that most mammals have. A stiff, brushlike mane runs the length of the neck. On its head, the giraffe has two short horns and sometimes two or three knobs. The horns and knobs are covered with skin.

Legs spread and head down, a giraffe stretches to reach the water. This awkward position makes the animal easy prey for big cats. Giraffes do not drink often. They get much of the moisture they need from their food.

Head high in a treetop, a giraffe feeds on the leaves of a thorny acacia. The animal avoids the largest thorns, and small ones cannot prick its tough lips.

Most giraffes have tan coats with irregular dark brown spots. But the reticulated (say rih-TICK-you-lay-ted) giraffe has large brown spots outlined in white. No two giraffes have the same pattern, but all giraffes have larger spots on their bodies than they do on their heads and legs. Giraffes from different areas have different kinds of patterns.

Some scientists think that a giraffe's spots may help hide the animal when it is standing in a shadowy grove of trees. The spots blend in with the shadows. And the giraffe's legs look like tree trunks. Its head is hidden in the leaves.

Even in the open, a giraffe has ways to protect itself from such enemies as lions. Its good eyesight allows the animal to see for long distances. And its height helps it see things that are far away, just as a person can see better from a lookout tower.

When a giraffe spots an enemy, it often has a good chance to gallop away before the attacker gets too close. A giraffe can run as fast as 35 miles (56 km) an hour. But it can't keep up that speed for very

Two bulls slam their necks together (below, left) in a "necking" bout that tests each animal's strength. They shove and push (below, center). And sometimes they even wrap their necks together as they spar (below, right). The fighting may look dangerous, but the bulls seldom get badly hurt.

long. Usually a giraffe lopes along at about 10 miles (16 km) an hour. Young giraffes can run faster than their parents because they weigh less. Adult giraffes can weigh as much as 2,800 pounds (1,270 kg).

Most of a giraffe's day is spent eating. Like a cow, a giraffe first swallows its food and then later brings it up to chew in the form of a cud—a wad of partly digested food. A giraffe has a four-chambered stomach like that of a cow.

Giraffes chew their cuds for hours at a time. The rest of the day, they wander and doze briefly. Giraffes nap standing up. Sometimes they lie down at night, but they never sleep deeply for more than a few minutes at a time.

After a 15-month pregnancy, a female giraffe gives birth—while standing—to one young. The offspring drops more than 5 feet (152 cm) to the ground when it is born! Then its mother turns and licks it. About half an hour later, the young giraffe can stand on its long, wobbly legs. It begins to

Necks swaying back and forth as they run, a small herd of young and adult giraffes lopes across a dry plain in Kenya. Giraffes can run as fast as 35 miles (56 km) an hour for short distances.

Giraffes at a water hole take turns drinking and watching for danger.

nurse. After only about ten hours, it can run alongside its mother. The newborn giraffe is tall—6 feet (183 cm)—and it weighs about 150 pounds (68 kg). It grows fast during its first year, adding almost 4 feet (122 cm) to its height. A giraffe is full grown after four or five years.

For many years, it was thought that giraffes were completely silent. But by watching and listening, scientists have now learned that the animals do have voices—very quiet ones.

GIRAFFE

HEIGHT: 14-19 ft (4-6 m)

WEIGHT: 1,750-2,800 lb (794-1,270 kg)

HABITAT AND RANGE: grasslands in central, eastern, and southern Africa

FOOD: leaves, twigs, and bark, especially from acacia trees

LIFE SPAN: about 25 years in the wild

REPRODUCTION: usually 1 young after a pregnancy of 15 months

ORDER: artiodactyls

Full-grown female licks an infant giraffe as a younger female stands by. Coats of brown spots outlined with thin bands of white identify these animals as reticulated giraffes.

Reticulated giraffe: 17 ft (5 m) tall

231

Gnu
Gnu is another name for wildebeest. Read about this animal on page 577.

Goat

AT HOME in rocky terrain, wild goats roam some of the most rugged mountains in the world. They scramble through barren, bone-dry mountains of the Middle East. In Europe, they climb steep cliffs and peaks. In Asia, on cold and windy slopes of the Himalayas, they range where few trees dot the landscape.

In North America, mountain goats live in a rocky, snowy world—high in the Rockies and in the mountains of the northwestern coast. Mountain goats are not really goats. But these shaggy-haired, surefooted climbers are close relatives. They are called goat-antelopes.

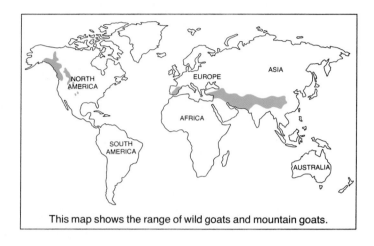

This map shows the range of wild goats and mountain goats.

Thick white coat protects a female mountain goat, or nanny, from icy winds high in the Rocky Mountains. With her split hooves and padded toes, she can keep her balance on snowy slopes.

Mountain goat: 42 in (107 cm) tall at the shoulder

232

Perched at the very edge of a cliff, two male mountain goats, called billies, rest in the sun. Mountain goats live in high country from Alaska through western Canada into the United States.

Goat

The hooves of mountain goats each are split into two toes. Rough pads on the bottom grip the surface of the ground. For an even better hold on slick slopes, a mountain goat's toes can spread wide.

A mountain goat's long outer coat protects it from wind, rain, and snow. This coat is shed in summer, leaving a layer of thick, fluffy wool. Both males, called billies, and females, called nannies, have beards and sharp black horns.

During most of the year, nannies and their young, called kids, travel in herds of as many as twenty animals. Billies live alone or occasionally in groups of two or three. They join the nannies and young at mating time. Then rival billies threaten each other. If a serious fight develops, they will lunge at each other's rumps with their daggerlike horns. Toward a female, a male acts differently. He crawls on his belly and bleats softly like a kid.

Young are born in late spring. A nanny usually

Balanced on a steep slope, a nanny feeds on tender new ▷ *leaves that cover a hillside. Her days-old kid stays close to her. A billy (in the small picture) nibbles tiny plants from the surface of a rock.*

▽ *Bath time for a billy means a roll in the dust. The spray of soil helps keep his coat free of oil and insects. Powerful shoulders and short, stocky legs give a mountain goat the strength a good climber needs.*

Toggenburg goat: 32 in (81 cm) tall at the shoulder

◁ *Mountain goats scramble at a dizzying height above Lake Ellen Wilson in Glacier National Park in Montana.*

▽ *Like many other domestic—or tame—goats, this nanny has no horns. Goats provide people with milk, cheese, and meat. And for centuries people have prized the fine wool of Angora and Kashmir goats.*

Mixed-breed domestic goat: 32 in (81 cm) tall at the shoulder

△ *Sleek and healthy, a Toggenburg goat of Switzerland grazes in a meadow. Swiss goats are champion milk producers. Easy and inexpensive to raise, goats are often called "the poor man's cow."*

has one kid. But she may have twins. Within minutes, a newborn kid struggles to its feet. Soon it is nursing, guzzling its first meal. Before long, it may try climbing its first steep slope.

Nannies keep careful watch over their young. Kids are frisky and often get into trouble. They might scramble up on a boulder and leap off. A frightened kid may run to its nanny for safety. The kid tucks itself between its mother's legs.

Kids stay with their mothers for almost a year—longer if she doesn't have another kid. During that time, they learn where to find the tender plants they eat. They learn where to seek shelter from storms. They learn where bears or wolves might prowl. They learn how to survive in their rugged world.

No real goats are native to North or South America. The common wild goat lives in dry, rocky parts of Asia and on islands in the Mediterranean. The Spanish goat lives in southern Spain. The markhor (say MAR-kor) and the tur (say TOUR) roam the high mountains of the southwestern Soviet Union and parts of Asia.

Goats were domesticated, or tamed, in what is now Iran at least 10,000 years ago. Today domestic goats provide people throughout the world with milk, cheese, meat, leather, and fine wool.

Goats can live in steep, bare places where domestic sheep and cattle cannot roam or find enough to eat. Often nothing grows in such areas because

GOAT

HEIGHT: 17-42 in (43-107 cm) at the shoulder

WEIGHT: 26-280 lb (12-127 kg)

HABITAT AND RANGE: mountainous regions of western North America and parts of Europe and Asia; domesticated goats live in many parts of the world

FOOD: grasses, herbs, trees, shrubs, and other plants

LIFE SPAN: 9 to 12 years in the wild

REPRODUCTION: 1 to 3 young after a pregnancy of 5 or 6 months

ORDER: artiodactyls

goats have already eaten all the plants. Tame or wild, goats feed on almost anything that grows. They nibble plants down to the ground and even uproot them. They snap twigs off bushes and peel bark off trees. Goats do not chew their food completely before swallowing it. After eating, they bring up a cud—a wad of partly digested food—and chew it thoroughly. Then they swallow and digest it.

Goats are closely related to sheep. One common difference between the two is the shape of their horns. A sheep's horns grow out to the side, then down and up again. A goat's horns usually grow straight up and then curve back.

The horns of the Kashmir markhor sweep up in graceful spirals. The Turkmen markhor's horns twist up in tight corkscrew curls. The tur and the bharal (say BUH-rul) have thick, rounded horns. Male wild goats have much larger horns than females have. They may also have flowing beards and manes. In addition, male goats give off a strong odor.

At mating time, the males fight with each other and show off

for the females. Wild goats fight head to head. Sometimes two bucks lock horns and pull sideways.

To learn more about goats and their relatives, read about chamois, ibexes, serows, sheep, and tahrs under their own headings.

Female Kashmir markhor leads the way across a rocky ▷ slope in Pakistan. Large, spiraling horns identify the male in the middle. Twin kids bring up the rear.

▽ Bharal, or blue sheep, bask in winter sunshine high above the tree line in Nepal. Despite their name and appearance, they act more like goats than like sheep. The heavy horns of male bharal keep growing throughout their lives. The bigger the horns, the older—and probably the stronger—the animal.

Bharal: 36 in (91 cm) tall at the shoulder

Kashmir markhor: 40 in (102 cm) tall at the shoulder

Turkmen markhor: 40 in (102 cm) tall at the shoulder

△ *Elegant swirling horns crown a regal Turkmen markhor. Its magnificent flowing mane sweeps down from its neck. Such goats make their homes in the high mountains of the southern Soviet Union.*

▽ *Popping out from behind a rock, a Spanish goat shows its impressive horns. Hunters have long prized as trophies the curving horns of this now rare animal.*

Spanish goat: 28 in (71 cm) tall at the shoulder

239

Gorilla

◁ Face framed by leaves, a gorilla stares from a tropical thicket in Africa. The crest of hair on its huge head shows that this animal is an older male.
Traveling on all fours, a group ▷ of gorillas moves from one lush feeding spot to another. The forest offers food at their fingertips —from roots and bamboo shoots to the bark and pulp of trees.
▽ Young gorilla rides on its mother's back through dense underbrush. A gorilla begins to travel in this way at about four months of age. Some continue for two years or more.

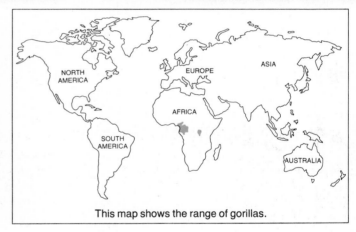

This map shows the range of gorillas.

"HU-HU-HU." Standing up and thumping his chest, an adult male gorilla can be an incredible sight. But don't let his looks scare you. The huge ape is quite peaceful most of the time.

A full-grown male gorilla can measure 6 feet (183 cm) long and weigh as much as 400 pounds (181 kg). Females are smaller and weigh only about half as much. Like the other apes—chimpanzees, orangutans, and gibbons—gorillas have no tails.

Gorillas roam the dense, moist forests of Africa. There they live in groups of two to thirty members. One kind of gorilla—the lowland gorilla—is found in west-central Africa. Scientists have not studied its habits in the wild. Another kind—the mountain gorilla—makes its home in forests in east-central Africa. Though they sometimes climb trees, adult gorillas spend most of the time on the ground. They usually travel along on all fours with their weight on their feet and on the knuckles of their hands. They can walk upright for short distances.

A mature male guides each group of mountain gorillas. He is called a silverback because he is older and the hair across his lower back has turned a silver color. The other gorillas follow his lead when it is time to feed, to travel, or to build night nests.

Silverbacks sometimes show off by hooting, standing up, and wildly throwing plants around. They beat their chests, (Continued on page 244)

241

Gorilla _____

Broad-shouldered male gorilla—called a silverback—surveys his range in a mountain forest in east-central Africa. He gets his name from the saddle of silver-colored hair across his lower back. A silverback often acts as the leader of a gorilla group. He guides other gorillas in their daily activities, determining when a group will move and when it will eat. The others also follow his example in building leafy nests in which to sleep. Sometimes, in a show of leadership, a silverback will stand upright and beat his chest. At right, a frisky young gorilla plays in a tree. Between the ages of three and six, gorillas often slide down trunks and climb and swing on branches.

△ *Soft tree pulp makes a snack for a female gorilla. Behind her, other gorillas rest in the underbrush.*

Fuzzy young gorilla plays with a leaf. Gorillas start ▷ eating plants several months after birth.

making hollow sounds that can be heard far away. The display shows a silverback's position as leader. It also is used to scare away intruders.

Gorillas have no trouble finding food in the lush forests where they live. Because of their size, these animals need a large amount of food every day. Their diet includes wild celery, roots, and the pulp and bark of trees. They also eat bamboo shoots and fruit in certain seasons.

After feeding, a group of gorillas settles down for a rest. Occasionally the animals pile together leafy plants for a day nest. Later the group again searches for food. As evening approaches, the gorillas make night nests, usually on the ground.

A female gorilla gives birth to one young after a pregnancy of about eight and a half months. The newborn is helpless and tiny, weighing less than 4 pounds (2 kg) at birth. At first, the young clings to its mother's chest. Later, it learns to ride on her back. By two years of age, a gorilla has begun to move

about on its own. But it may share its mother's nest until it is five.

Between the ages of three and six, gorillas act much like children. They spend their time playing. They climb trees, swing on branches, slide down tree trunks, chase each other through the forest, and even play a game that looks like tug-of-war. They also imitate adult gorillas.

GORILLA

LENGTH OF HEAD AND BODY: 4-6 ft (122-183 cm)

WEIGHT: 150-400 lb (68-181 kg)

HABITAT AND RANGE: dense, moist lowland and mountain forests in west-central and east-central Africa

FOOD: wild celery, roots, tree bark and pulp, fruit, and bamboo shoots

LIFE SPAN: 35 years in captivity

REPRODUCTION: usually 1 young after a pregnancy of about 8½ months

ORDER: primates

◁ *Dangling in midair, a female gorilla hangs by her long, strong arms. As gorillas jump down from trees, they may break branches simply to make noise.*

▽ *Mossy fork of a massive tree provides a comfortable sun deck for an adult gorilla. More often, adults sunbathe on the ground. Gorillas frequently bask for an hour or two at a time.*

Lesser grison: 17 in (43 cm) long; tail, 7 in (18 cm)

Grison

(say GRIZ-un)

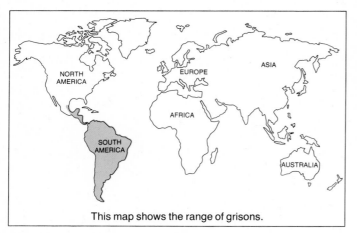
This map shows the range of grisons.

Common grison: 20 in (51 cm) long; tail, 7 in (18 cm)

Headband of white fur highlights a common grison (above) as it pokes its head out of a hole. At top, a lesser grison sits in a tree. These animals usually search for food on the ground.

GRISON

LENGTH OF HEAD AND BODY: 16-23 in (41-58 cm); tail, 6-8 in (15-20 cm)

WEIGHT: 2-7 lb (1-3 kg)

HABITAT AND RANGE: tropical forests, woodlands, and grasslands from Mexico through South America

FOOD: rodents, young birds, eggs, and fruit

LIFE SPAN: 10 years in captivity

REPRODUCTION: 2 to 4 young after a pregnancy of unknown length

ORDER: carnivores

A HOLLOW LOG or a rocky shelter may serve as a den for the long, slender grison of Mexico and Central and South America. Or this member of the weasel family may take over a large rodent's burrow. Grisons make their homes in forests, on grasslands, and sometimes near towns. Scientists think grisons live alone, except when raising young. There are usually two to four offspring in a litter.

Moving briskly along the ground, grisons hunt

both by day and by night. They feed on small animals. When a grison goes after a rodent like a vizcacha, it may follow the prey into its burrow. Grisons also catch birds on the ground.

Like its relative the badger, the grison has a black-and-gray coat with white markings. A white stripe goes across its forehead and down to its shoulders. Grisons also have scent glands. These give off a strong-smelling odor when the animals are excited or threatened. Read about badgers and other relatives of grisons—ratels, skunks, and zorillas—under their own headings.

Groundhog

Groundhog is another name for the woodchuck. Learn about woodchucks and other marmots on page 358.

Guanaco

The guanaco is a close relative of the llama. Read about both animals on page 342.

Guenon

The guenon is a kind of monkey. Read about guenons and other monkeys on page 376.

Guinea pig

(*say* GINN-ee pig)

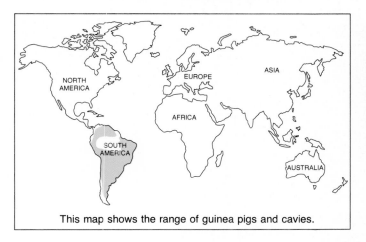

This map shows the range of guinea pigs and cavies.

Long-haired guinea pig: 10 in (25 cm) long

Short-haired guinea pig: 10 in (25 cm) long

ONCE YOU HAVE HEARD the sound it makes, you can guess why this little animal has "pig" in its name. When the guinea pig gets excited, it often squeals like a pig.

The guinea pig is not related to the pig. This

Guinea pigs may differ greatly in the kinds of coats they have. A short-haired guinea pig (right) climbs over toadstools in its path. The coat of the multicolored long-haired guinea pig (top, right) sweeps the ground. Although one guinea pig may look very different from another, there is only one kind, or species, of guinea pig.

plump, short-legged animal is really a rodent. Guinea pigs are much smaller than real pigs. They measure only about 10 inches (25 cm) long.

A guinea pig may be black, white, yellow, brown, or a combination of these colors. Its hair may be short or long, straight or curly. Though one guinea pig may differ from another, there is only one species, or kind, of guinea pig.

Female guinea pigs usually bear one to four offspring after a pregnancy of about two months. Young guinea pigs nurse for about three weeks. But, if necessary, they can survive without their mothers after about five days.

The guinea pig developed from a small wild animal called the cavy (say KAY-vee). Many hundreds of years ago, people in South America began to capture cavies and to raise them for food. Explorers saw these animals and took some of them back to Europe. After years of breeding, the wild cavy became the tame guinea pig we know today.

GUINEA PIG AND CAVY

LENGTH OF HEAD AND BODY: 6-16 in (15-41 cm)

WEIGHT: 11-50 oz (312-1,418 g)

HABITAT AND RANGE: rocky areas, grasslands, open woodlands, swamps, and dry plains in parts of South America; guinea pigs are kept as pets in many parts of the world

FOOD: grasses, herbs, and other plants

LIFE SPAN: about 8 years in captivity

REPRODUCTION: usually 1 to 4 young after a pregnancy of about 2 months

ORDER: rodents

Desert cavy: 8 in (20 cm) long

△ *Wild relative of the guinea pig, a desert cavy stretches to pluck a berry from a bush. These cavies live on dry plains in southern Argentina. The color of their coats blends well with the sandy soil.*

◁ *Female desert cavy nurses her young, even though they can eat some plants soon after birth. These newborn—like newborn guinea pigs—have hair and teeth. A desert cavy usually gives birth to two young in a litter.*

The guinea pig still has many wild relatives in South America. All of them are members of the cavy family. Of about fifteen species of cavies, most are the same size as the guinea pig. The desert cavy, for instance, measures only about 8 inches (20 cm) long. It lives on dry plains in southern Argentina.

Other species of cavies live in different parts of South America and in many kinds of terrain—rocky areas, grasslands, open woodlands, and swamps. Cavies, like many rodents, eat grasses, herbs, and other plants. They have sharp front teeth for cutting and gnawing and strong back teeth for grinding. Their teeth grow all their lives. Unless a rodent gnaws a great deal, its front teeth become so long that it cannot close its mouth. Then it is not able to chew or to swallow its food.

Gymnure

(say JIM-nur)

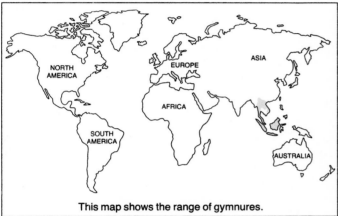

This map shows the range of gymnures.

Moonrat: 17 in (43 cm) long; tail, 8 in (20 cm)

Pointed teeth and bristling fur warn intruders to leave this gymnure—called a moonrat—alone. When threatened, the moonrat, largest of several kinds of gymnures, opens its mouth wide and hisses loudly.

WADDLING ALONG THE FOREST FLOOR at night, the gymnure noses through leaves and underbrush, searching for worms, insects, and snails to eat. There are several kinds of gymnures. One kind, the moonrat, lives near water and catches frogs and fish. As a gymnure travels about, glands at the base of its tail produce an unpleasant-smelling substance. The gymnure leaves this scent along its path. At dawn, it curls up in a hollow log to sleep.

With its small round ears, whiskered snout, and nearly hairless tail, the gymnure looks almost like a rodent. But it is not—it is in the same family as the hedgehog. Because it has a fuzzy coat, this animal of Southeast Asia is often called a hairy hedgehog. Find out about hedgehogs on page 256.

The gymnure was not known to scientists until about 150 years ago. Although experts now have identified at least four kinds of gymnures, they still know little about the animals.

GYMNURE

LENGTH OF HEAD AND BODY: 4-17 in (10-43 cm); tail, 1-8 in (3-20 cm)

WEIGHT: 2-49 oz (57-1,389 g)

HABITAT AND RANGE: forests and mangrove swamps in Southeast Asia

FOOD: insects, worms, plants, and small water animals

LIFE SPAN: more than 7 years in captivity for moonrat; unknown for other species

REPRODUCTION: unknown

ORDER: insectivores

249

H

Hamster

(*say* HAM-stir)

THOUGH IT IS FAMILIAR as a household pet, the hamster also lives wild in parts of Europe and Asia. There this rat-size rodent digs burrows along riverbanks, in desert areas, in fields, and sometimes on mountain slopes. These underground homes usually have storerooms, a nest area, and several entrances. There is even a separate area that serves as a toilet.

Hamsters range in length from 2 to 11 inches (5-28 cm). They usually have very short tails, but a few kinds have longer ones. Most hamsters have large pouches in their cheeks. They use these to carry food back to their burrows.

Golden hamster: 7 in (18 cm) long

Female hamsters give birth several times a year. The young—usually 4 to 12 in each litter—are born blind and helpless. They grow to full size quickly.

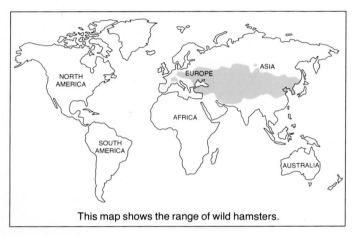

This map shows the range of wild hamsters.

HAMSTER

LENGTH OF HEAD AND BODY: 2-11 in (5-28 cm); tail, as long as 4 in (10 cm)

WEIGHT: as much as 32 oz (907 g)

HABITAT AND RANGE: fields, riverbanks, mountains, and desert areas in parts of Europe and Asia; hamsters are kept as pets in some parts of the world

FOOD: fruit, grain, seeds, roots, and small animals

LIFE SPAN: 2 or 3 years in captivity

REPRODUCTION: usually 4 to 12 young after a pregnancy of 2 or 3 weeks

ORDER: rodents

Shiny black eyes of a golden hamster stare out from the animal's round face. Cheek pouches stuffed with food give this hamster a well-fed look.

Hare

FROM A STANDSTILL, the hare can explode in a powerful jump straight into the air. This furry, long-eared animal can run at speeds of about 40 miles (64 km) an hour. Some hares can also leap forward as far as 10 feet (305 cm) with their long legs and large hind feet. These are impressive feats for an animal that

measures only about 2 feet (61 cm) long from twitchy nose to stubby tail.

Most people cannot tell the difference between a hare and a rabbit. The animals are close relatives, and both live almost everywhere in the world. Hares and rabbits look very much alike. Even their names

Stretching its powerful hind legs forward, a white-tailed jackrabbit races across a snow-covered field.

are confusing. A jackrabbit is really a hare. A Belgian hare is really a rabbit. Usually a hare is larger than a rabbit, and it has longer hind legs. The ears of a hare are also longer than those of a rabbit and are often tipped with black.

There are more than twenty kinds of hares. They can survive in almost any climate and in any terrain, as long as they can find shelter and low-growing plants to eat. The animals live in open country and in forests, in mountains and in deserts.

In North America, most hares live alone. They generally do not live in burrows. Their dens, called forms, are often in the open and are not much more than shallow places in the ground. A hare digs this kind of resting place with its forefeet. A single hare may have several forms. Some may use the empty dens of foxes and marmots. Others may seek shelter in caves or under rocks.

A hare's life can be very dangerous. Coyotes, lynxes, foxes, eagles, hawks, and owls prey on hares. Hares are hunted by people, too. For centuries, they have killed hares for their meat, for their fur, and for sport. The animals can be pests to farmers, eating garden vegetables and other crops.

HARE

LENGTH OF HEAD AND BODY: **14-28 in (36-71 cm); tail, 2-4 in (5-10 cm)**

WEIGHT: **3-12 lb (1-5 kg)**

HABITAT AND RANGE: **forests, grasslands, tundra, deserts, and mountain slopes worldwide, except for Antarctica and some oceanic islands.**

FOOD: **grasses, herbs, shrubs, twigs, and bark**

LIFE SPAN: **as long as 7 years in captivity**

REPRODUCTION: **2 to 8 young after a pregnancy of about 1½ months**

ORDER: **lagomorphs**

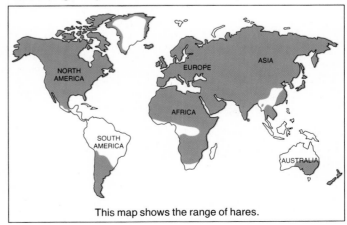

This map shows the range of hares.

Blue hare: 21 in (53 cm) long; tail, 3 in (8 cm)

Most kinds of hares are in no danger of dying out, however, because there are so many of them. Some female hares have as many as four litters a year. Each litter may include from two to eight young. A newborn hare, called a leveret (say LEV-uh-rut), has its eyes open and is covered with fur. It can hop just minutes after it is born. A female hare spends little time taking care of her young.

Hares eat only plants. They feed alone, usually in the morning and in the evening. The animals nibble mainly grasses and herbs. When food is scarce in winter, they eat shrubs, twigs, and even bark. The little water they need comes mostly from the plants they eat. Hares cut or nip plants with their sharp front teeth. Hares grow two pairs of upper front teeth, one set right behind the other. The front pair continues to grow as long as the animals live, just as a rodent's front teeth do. Hares are not rodents, though. Scientists put them in a separate group with pikas and rabbits. Read about pikas on page 442 and rabbits on page 472.

As they hop around their home range, hares often mark the land with scent. They even mark

◁ *Protected by its coloring, a blue hare pauses at the entrance of its resting place high in the Alps of Switzerland. When the snow melts, this hare's coat will change to brown.*

▽ *From winter to summer, the color of a snowshoe hare gradually changes. In winter, the showshoe hare's fur is light colored (below, left) and grows in long and thick. In late spring (below, center), its fur coat shows a mix of white and brown. A hare's summer coat (below, right) grows in darker and shorter. Smallest of the more than twenty kinds of hares—a showshoe hare weighs only about 3 pounds (1 kg).*

Snowshoe hare: 15 in (38 cm) long; tail, 2 in (5 cm)

253

Hare

Cape hare freezes, ▷ sitting perfectly still among plants in South Africa. The lack of movement helps hide the animal from its enemies. Eyes on the sides of its head give a hare a wide angle of vision. It sees best at dusk.

Cape hare: 22 in (56 cm) long; tail, 4 in (10 cm)

Black-naped hare: 14 in (36 cm) long; tail, 3 in (8 cm)

Antelope jackrabbit: 21 in (53 cm) long; tail, 3 in (8 cm)

△ Black-naped hare, named for the dark patch of fur on the back of its neck, feeds on grasses in Sri Lanka. This kind of hare may take shelter in caves or in hollow logs.
◁ Antelope jackrabbit skims the ground with all four feet in the air. These large hares, the fastest of all, can reach speeds of 40 miles (64 km) an hour. They may cover 10 feet (305 cm) in a single leap. Their ears measure about 7 inches (18 cm) long— about one-third the length of their bodies.

Young European hares, barely one week old (below, left), nestle together for warmth in an English marsh. Two adults (below, right) stand ready to fight during the mating season.

European hare: 22 in (56 cm) long; tail, 4 in (10 cm)

Arctic hare: 21 in (53 cm) long; tail, 2 in (5 cm)

◁ *Well camouflaged in its coat of white, an arctic hare in northern Canada fluffs up its fur for warmth. In the Far North, hares stay white all year. Farther south, some arctic hares change to a grayish color in summer.*

▽ *In early evening, a black-tailed jackrabbit sniffs for food or for the scent of other hares. This North American hare often takes shelter near bushes.*

themselves. Special glands on a hare's lower jaw put out a strong-smelling liquid. A hare rubs its jaw on the ground or on a twig to mark the area as its own. When a hare grooms itself, it spreads some of this liquid over its body. Because of other scent glands near its tail, a hare can leave its scent while sitting. This strong smell lets other hares know that it has been there. Scent also helps males and females find each other at mating time.

To protect themselves, hares rely on their senses of hearing, smell, and sight. When its senses alert it to danger, a hare may react in one of several ways. If the enemy is far away, the hare may sneak away, keeping its ears low and running close to the ground. It hides among nearby shrubs. If an enemy is close by, though, the hare may freeze. It sits still, head down, ears back, and nose twitching. It is difficult to see the hare—even on bare ground. If the enemy comes closer, the hare may pop into the air in a mighty leap. It dashes away with ears up and tail down. The hare can outrun many animals that chase it.

When a hare is being chased, it tries to stay within its home range. At first, it runs in a straight line. If the enemy gets too close, however, the hare may run in a zigzag pattern. Running zigzag helps throw its pursuer off balance. A hare may even swim a short distance to escape.

A hare runs by moving its forefeet first. Then it jerks its rear end forward. Its hind feet hit the ground in front of its forefeet. When moving slowly, a hare curves its tail along its back. When the animal runs fast, its tail moves up and down.

Black-tailed jackrabbit: 19 in (48 cm) long; tail, 3 in (8 cm)

All hares molt, or gradually shed their hair. Some molt twice a year. For most of these animals, summer hair is short and dark. Winter hair is usually thick, long, and light. Some North American hares grow white hair in autumn. This coloring helps hide them during snowy winters. The snowshoe hare changes from white in winter to grayish brown in summer. The change explains its other name, varying hare. Most arctic hares stay white all year round, but a few molt to a grayish color in summer. Hares in desert areas are almost red. Some hares in Asia are completely black.

Hares have lived near people for centuries. They appear in stories and folklore around the world. More than 2,000 years ago, the Greek storyteller Aesop created the fable of the speedy hare that lost a race to the slow-but-steady tortoise. And one of the most famous characters in *Alice in Wonderland* is the March Hare.

Hartebeest

The hartebeest is a kind of antelope. Read about antelopes on page 52.

Hedgehog

(say HEDGE-hog)

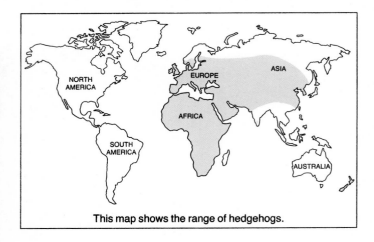

This map shows the range of hedgehogs.

HEDGEHOG

LENGTH OF HEAD AND BODY: 5-12 in (13-30 cm); tail, 1-2 in (3-5 cm)

WEIGHT: 14-39 oz (397-1,106 g)

HABITAT AND RANGE: forests, plains, and deserts of Europe, Asia, Africa, and New Zealand

FOOD: insects, snails, mice, birds, frogs, lizards, and snakes

LIFE SPAN: about 10 years in captivity

REPRODUCTION: 1 to 7 young after a pregnancy of 1 or 2 months, depending on species

ORDER: insectivores

European hedgehog: 11 in (28 cm) long; tail, 1 in (3 cm)

△ *Soft, white spines stick out from the backs of five-day-old hedgehogs in a grassy nest. As the animals grow, their spines will become stiffer, sharper, and longer. At about two weeks of age, young hedgehogs open their eyes. Soon after, they begin to follow their mother as she hunts for food. After several months, they can live on their own. A full-grown European hedgehog (top) sits by a mountain stream before jumping in.*

SMALL ENOUGH TO FIT IN YOUR HANDS—but too prickly to hold—the hedgehog has armor that resembles needles in a pincushion. Thousands of stiff, sharp spines stick out from the animal's back.

By curling up and tucking in its head and legs, the hedgehog can turn its body into a spiked ball. A rounded shield of spines almost completely encloses its hairy face and underparts. In this position, the hedgehog is usually safe from attackers. It even sleeps curled up.

Only a few animals—such as badgers and foxes—can pry open a rolled-up hedgehog. Because most other animals leave it alone, the hedgehog does not need to keep silent to avoid discovery. When angry, it screams. It often snorts, coughs, and wheezes as it pokes about in the dirt for food.

Hedgehogs feed on cockroaches, snails, young mice, birds, frogs, and lizards. Some people keep hedgehogs as pets because they eat garden pests. Hedgehogs also feed on bees and wasps. The stings seem to have no effect on the hardy little mammals. Sometimes a hedgehog will even attack a poisonous snake for food. Usually its spiny coat will protect it from the fangs.

Hedgehogs can live in many kinds of climates and terrains. They are found in parts of Europe, Africa, and Asia. Settlers took them to New Zealand about a hundred years ago.

A hedgehog roams within its own small territory. Each night, it trots along the same pathways looking for food. It sleeps during the day in a grass-lined nest hidden by rocks, hedges, or underbrush.

If food is scarce, a hedgehog may sleep for weeks at a time. In cool climates, the animal's long sleep takes place in winter. Its breathing and heart rate slow down, and its temperature falls. Scientists call this hibernation (say hye-bur-NAY-shun).

In desert regions, a hedgehog sleeps through the driest summer weather. This hot-weather slumber is called aestivation (say es-tuh-VAY-shun). An animal that aestivates does not sleep as deeply as one that hibernates.

Female hedgehogs give birth to as many as seven young in a litter. The newborn are covered with soft, white spines. Their spines become stiffer, sharper, and longer as the offspring grow.

African hedgehog: 10 in (25 cm) long; tail, 1 in (3 cm)

◁ *African hedgehog scampers across sandy soil hunting for food. Usually it stays in its nest during the day.*

△ *Prickly as a pincushion full of needles, an African hedgehog rolls into a tight ball to defend itself.*

◁ *Wiggling snake makes a meal for a long-eared hedgehog. The hedgehog's spiny armor usually protects it from bites.*

Long-eared hedgehog: 9 in (23 cm) long; tail, 1 in (3 cm)

257

Hippopotamus

(*say* hip-uh-POT-uh-muss)

△ *Ears, eyes, and nostrils above water, a bull river hippo rests its muzzle on another's back. In the water, a hippo usually keeps most of its body under the surface.*

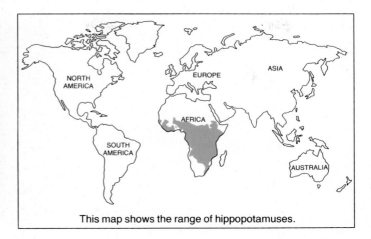

This map shows the range of hippopotamuses.

DESPITE ITS SIZE and awkward looks, the river hippopotamus moves lightly in water. This huge animal walks like a dancer in slow motion on the bottom of a calm river. As it rises to take a breath, its great head breaks the surface. Bulging eyes pop open. Ears unfold and wag, sending up a shower of water droplets. "Un-n-nk," grunts the hippo loudly. Other hippos answer.

The hippopotamus is one of the largest land mammals in the world. A bull hippo can measure 15 feet (5 m) long and can weigh 8,000 pounds (3,629 kg). A female is usually smaller.

People have known about the hippopotamus for centuries. Ancient Egyptians made models of hippos and glazed them blue. Greeks named the animal *hippopotamus,* which means "river horse." Hippos are more closely related to pigs than to horses, however. Romans exhibited captured hippos in their arenas.

Until recent times, hippos ranged throughout most of Africa. They could be seen in lakes, streams, and rivers. Because of hunting and the spread of farms, hippos have now disappeared from many areas. Today these barrel-shaped animals live mostly in central and southern Africa. In a few places, there still may be as many as 2,000 hippos in a 20-mile (32-km) stretch of a river.

The hippo is well adapted, or suited, to its life in the water. With eyes, ears, and nostrils on the top of its head, it can see, hear, and breathe even when most of its body remains underwater. A good swimmer, the animal stays in water much of the day.

From time to time, a hippo may sink from sight beneath the surface. An adult hippo may stay on the bottom for as long as five minutes. Its heart rate slows down. The animal may walk along the bottom, following underwater trails that it and other hippos have made. The hippo's great weight keeps it from floating to the surface.

A hippo's skin is thick and almost hairless. Oily red drops ooze from its pores. People once thought that the animal was sweating blood. But this thick,

◁ *Female hippos and young rest in the calm waters of an African river. Some sun themselves on the sandy bank. Others cool off in the shallows at the river's edge.*

HIPPOPOTAMUS

LENGTH OF HEAD AND BODY: **12-15 ft (4-5 m); tail, 16-22 in (41-56 cm). Pygmy: 59-69 in (150-175 cm); tail, 6-7 in (15-18 cm)**

WEIGHT: **5,000-8,000 lb (2,268-3,629 kg). Pygmy: 400-600 lb (181-272 kg)**

HABITAT AND RANGE: **rivers, lakes, and streams of central and southern Africa. Pygmy: wet forests in parts of western Africa**

FOOD: **grasses and water plants. Pygmy: leaves, shoots, grasses, and fruit**

LIFE SPAN: **more than 40 years in the wild. Pygmy: as long as 38 years in captivity**

REPRODUCTION: **1 young after a pregnancy of 7 or 8 months, depending on species**

ORDER: **artiodactyls**

△ *Open w-i-d-e! A female hippo yawns, flashing pink gums and pointed tusks. An open mouth may mean excitement, or it may signal a threat. Hippos also open their mouths wide before they leave the water to feed.*

Rearing up with a splash and a roar, a male and a ▷ female hippo slam their jaws together in a courtship display. Other females surround them.

red liquid actually keeps the hippo's skin moist. It also may help to kill germs and to heal wounds.

After sundown—earlier on cloudy days—hippos leave the water to feed on land. A hippopotamus may look fat and awkward, but its body is muscular. Hippos easily climb steep banks using well-worn footholds. Out of the water, hippos walk in line on narrow paths to grazing grounds as far as 5 miles (8 km) away. If danger threatens, hippos will hurry back to the safety of the water. The animals can run as fast as a person can over short distances.

Although hippos sometimes eat water plants, grass is their main food. A hippo closes its huge, tough lips around a mouthful of grass, then tears the

food away neatly. As they munch the grass, hippos make a lot of noise. Mothers and their young feed close together. Others feed alone. An adult hippo may eat 150 pounds (68 kg) of grass in a night. That may seem like a lot, but it's not very much for such a huge animal.

A hippo's teeth continue to grow throughout the animal's life. The teeth used for chewing are worn down as the animal grows older. In its lower jaw are two tusks, which may grow more than 1 foot (30 cm) long. These tusks wear against a hippo's upper teeth and stay sharp.

Strong male hippos control territories, or

mating areas, in the water and on the nearby land. The territory can measure about 1,000 feet (305 m) long and 165 feet (50 m) wide. A bull hippo is master of his territory. But he allows other male hippos in as visitors. He may challenge them by opening his enormous mouth and displaying his tusks. The visitors must show their respect or he will chase them away. Sometimes two bull hippopotamuses challenge each other at the boundary of their territories. With mouths open, they rush at each other and meet jaw to jaw. The clash continues until one animal gives up and both retreat.

If one hippo tries to take over the territory of another, the two fight in a different way. Facing in opposite directions, the animals stand side by side. They swing their heads sideways and up. Each one tries to drive his sharp tusks into the other's sides and rump. The fights may last more than an hour. The blows strike where the hippo's hide is more than 2 inches (5 cm) thick, but the tusks still stab deeply. Scars mark the bodies of many bulls.

Hippos mate in shallow water. Young are born in the water or on land. After a pregnancy of eight months, a female gives birth to a single grayish pink calf that weighs 100 pounds (45 kg). Sometimes a calf nurses on land, lying like a piglet beside its

△ *Like islands in a stream, river hippos provide motionless perches for a group of snow-white cattle egrets. When the hippos leave the water for land, the birds follow and catch insects that fly up.*

◁ *Beneath an arch of plants, a young male hippo walks gracefully on the bottom of a shallow pool. Scientists think the fish swimming nearby clean the hippo's skin of bits of plants.*

mother. At other times it may nurse underwater. But in either case, the calf sucks with its nostrils closed and its ears folded shut. It must stop drinking from time to time to breathe.

A young hippo may scramble onto its mother's back and rest while she lies in the water. That way it does not have to struggle to stay afloat.

After a few days, a female hippo and her newborn join the other mothers and calves in a nursery herd. The little ones chase each other and play at fighting. Mothers will stay near the water to feed until the calves are strong enough to keep up on the trips to the grazing grounds. The female walks very close to her offspring. When danger threatens, she defends her young furiously.

A young hippo's first year is a dangerous time. Lions, leopards, hyenas, crocodiles, and wild dogs prey on small hippos. Adults are rarely attacked by other kinds of animals. Hippos may live for more than forty years in the wild.

Another kind of hippo lives in swampy forests in a small area of western Africa. The pygmy hippopotamus, a hog-size animal, is not simply a small copy of its huge relative. Its eyes are set in the sides of its head instead of on top. Its skin is kept moist by a clear liquid instead of a red one.

Not much is known about this rare animal in the wild, though its habits seem to be different from those of the river hippo. The pygmy hippo spends most of its time on land. There it travels along tunnel-like paths in search of food.

Pygmy hippos feed on leaves, shoots, grasses, and fruit. Scientists think that adult pygmy hippos live alone or perhaps in pairs. A calf weighs about 10 pounds (5 kg) at birth.

▽ *Pygmy hippo looks like a hog and weighs only about one-tenth as much as its larger relative. This animal spends most of its time on land rather than in the water. It feeds on leaves, shoots, grasses, and fruit.*

Pygmy hippo: 5 ft (152 cm) long; tail, 7 in (18 cm)

263

Hog

Wild boar sniffs the air in a forest in France. Despite its shaggy coat, the animal resembles its tame relatives. Adaptable animals, hogs can eat many kinds of food and live in many kinds of terrain.

IF YOU COULD CLIMB aboard a time machine and travel back millions of years, you might find a familiar animal: the hog. Scientists think that the hog has existed for at least 45 million years. And in all that time its stocky, round body and flat snout have changed very little. Hogs go by many names. They are known as swine and as pigs. Male hogs and some kinds of wild pigs are called boars. Females are sows. Young hogs are called piglets.

Domestic, or tame, hogs are important food animals on every continent except Antarctica. Wild pigs roam forests, meadows, and swamps around the world. Africa is home to the giant forest hog and the fierce-looking warthog. The bearded pig lives in Indonesia. In the southern foothills of the Himalayas in Asia, rust-brown pygmy hogs live in the underbrush. Wild boars are found in scattered parts of North America.

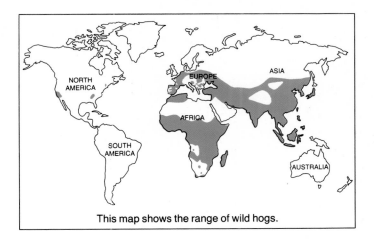

This map shows the range of wild hogs.

HOG

HEIGHT: **12-40 in (30-102 cm) at the shoulder**

WEIGHT: **13-800 lb (6-363 kg)**

HABITAT AND RANGE: **forests, grasslands, meadows, and swamps in Africa, Asia, Europe, and North America; domestic hogs live in many parts of the world**

FOOD: **small animals and all kinds of plants**

LIFE SPAN: **15 to 20 years in the wild**

REPRODUCTION: **2 to 12 young after a pregnancy of about 3½ to 5 months; domestic hogs may have many more young**

ORDER: **artiodactyls**

Hogs will eat just about anything they can get their snouts on: roots, grasses, fruit, nuts, herbs, mushrooms, worms, and even snakes. Hogs have been able to live successfully in so many different places because they can find food almost anywhere. To find food, the animals often root, or dig into the ground with their snouts. Pigs hardly ever overeat, even though we call a person who stuffs himself with food a "pig."

Because hogs like to wallow in mud, people sometimes think of them as dirty animals. But when they have a choice hogs stay in clean water, rather than in mud. Lying in water helps hogs cool off on a hot day. The animals have few sweat glands in their skin so they cannot cool off by sweating. They take frequent cool baths to keep their body temperatures low and to escape from insects.

Hogs are intelligent animals. Like dogs, they can learn to sit up and to roll over on command. Some hogs even appear in circus acts, dancing and doing tricks. Hogs used in experiments have learned how to switch on heat lamps to warm up their cages. Centuries ago, people in England trained hogs as hunting companions. Because of their sensitive noses and keen hearing, they made good helpers for the hunters.

In the wild, most pigs live in groups called sounders, although some older boars live alone. Hogs stay close to each other most of the time. Piglets pile close together when they sleep. Since they have little hair, this helps keep them warm.

Hogs communicate with each other by making different noises. High-pitched squeals signal danger. Low grunts mean contentment. Hogs often grunt quietly as they eat. The animals produce a variety of other sounds. They use them during courtship and while searching for food.

Hogs come in many sizes. The smallest member of the pig family, the pygmy hog, grows only about 1 foot (30 cm) tall at the shoulder. It weighs about 13 pounds (6 kg). The giant forest hog is about 40 inches (102 cm) tall and weighs 300 pounds (136 kg). Some domestic hogs are even heavier. Farmers fatten some of their breeding hogs until they weigh as much as 800 pounds (363 kg)!

Also large—and much hardier than domestic

Wild piglet about two weeks old stands unsteadily in a field in Malaysia. The light stripes on its coat will begin to fade within three months. After a year, the animal will grow thick, bristly hair.

Southeast Asian pig: 33 in (84 cm) tall at the shoulder when fully grown

265

Hog

▽ *Warthog sow and her piglet wallow in mud along a river in South Africa. Such mud baths keep the warthogs cool and help them escape from biting insects. Warthog piglets do not have striped coats like those of other young hogs. Instead, they have thick, reddish hair.*

Warthog: 30 in (76 cm) tall at the shoulder

Giant forest hog: 40 in (102 cm) tall at the shoulder

△ *Giant forest hogs, marked by crescent-shaped growths under their eyes, head for a water hole in Kenya. Scientists first discovered this kind of hog in 1904.*

hogs—are European wild boars. These bristly animals have massive bodies and large heads and can weigh as much as 700 pounds (318 kg). Males have long, sharp tusks, which they use in fights over females. But males have some protection from each other's tusks. Just before mating season, when the battles occur, the skin on boars' flanks grows especially thick. Female wild boars also have tusks. But they use them for digging rather than for fighting.

Wild boars wander through the forests of many European countries, including Germany, Spain, Belgium, Holland, and France. In England, they died out more than 200 years ago. Wild boars also live in many Asian countries. Domestic hogs are descendants of the wild boar.

The bushpig is another wild member of the pig family. Bushpigs live in most parts of Africa south of the Sahara. Like many wild swine, they are active after dark. They spend much time rooting in the ground for their food. These wild pigs grow about 30 inches (76 cm) tall, and they have short, very sharp tusks. Some kinds are colorful. The red river hog, for example, has bright white markings on its rust-colored coat.

The warthog is one of the fiercest-looking mammals in the world. The male's long, flat head is dotted with bumps that look like huge warts. Four pointed tusks stick out from the sides of its large snout. Its face is long, and tiny eyes shine from the top of its forehead.

In spite of its appearance, the warthog is a peaceful animal. Rather than fight, it will turn and run away from an enemy. With its tufted tail sticking straight up into the air, the animal gallops along at 30 miles (48 km) an hour.

A warthog often protects itself by hiding in an

◁ *Hightailing it across the grasslands of Africa, a group of warthogs runs for cover. The animals point their tufted tails straight into the air as they go. Warthogs can move as fast as 30 miles (48 km) an hour.*

▽ *Yellow-billed oxpecker perches on a warthog's shoulder. The bird searches for ticks on the skin of the fierce-looking animal. The oxpecker also may signal danger. If another animal comes near, the bird will fly off. This alerts the warthog to watch out for trouble.*

empty aardvark den. When a group of warthogs is in danger, the young scurry quickly into the den. Then an adult backs in, blocking the entrance and protecting the young behind it. Any animal that tries to follow runs right into the adult warthog's sharp tusks.

Female warthogs also use abandoned aardvark dens to raise their young. Other kinds of wild pigs prepare nests for their offspring. Before giving

Chester White hog: 36 in (91 cm) tall at the shoulder when fully grown

◁ *Three-week-old Chester White piglets snuggle for warmth on a hog farm. People throughout the United States raise these hogs for their meat.*

▽ *Colorful coat of a female red river hog in Africa shows why people sometimes call this animal "the dandy of the pig family."*

△ *Indian women in Ecuador herd hogs in a mountain*

birth, a pregnant sow will go off by herself to make the nest. She builds a mound of leaves and grasses and hides her young inside.

Both wild and domestic sows are careful about raising their young. A mother hog will fiercely defend her offspring from danger. She will fight anything that dares to approach her litter. A wild sow

Red river hog: 30 in (76 cm) tall at the shoulder

pasture. The dark woolly coats of the hogs protect them from cool temperatures and heavy rains.

gives birth to 2 to 12 piglets. Domestic hogs may have even larger litters: On a hog farm in the Midwest, one sow gave birth to 27 piglets.

Usually piglets in the wild try solid food soon after birth. But they continue to nurse for about three months. Even when they have become quite large, the piglets will chase their mother and butt against her side until she lets them drink her milk.

Pigs are familiar animals, not only because we see them on farms, but also because they turn up in stories, sayings, and rhymes. Just about everyone knows the tale of "The Three Little Pigs." And most people have had their feet tickled as someone recited, "This little piggy went to market. . . ."

Horse

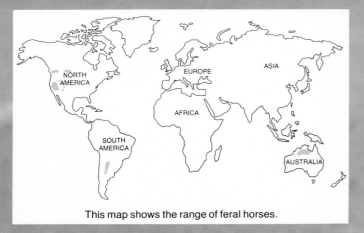

This map shows the range of feral horses.

MANE FLYING AND HOOVES THUNDERING, a galloping horse inspires images of wildness and freedom. For centuries, people have admired horses for their strength, their beauty, their spirit, and their speed. Horses probably were first tamed by nomadic peoples in Asia about 4,000 years ago. They captured wild horses with thick coats that roamed the treeless plains. They used the horses as work animals and drank the milk of the mares. Gradually, the practice of taming and breeding horses spread into Europe. From there, explorers brought the animals to North and South America.

Horses are well equipped for running. With

Mustang mares and their long-legged foals gallop across a rolling prairie in Wyoming. These wild descendants of tame horses roam in scattered bands in parts of the western United States.

Mustang: 55 in (140 cm) tall at the withers — the ridge between the shoulders

their strong, muscular bodies, they can run for many miles at a time. Their nostrils flare wide to let air into their large lungs. Though a horse pounds the ground with tremendous force as it runs, its feet are protected by hard hooves. The foot is actually a single toe, and the hoof is like a very thick toenail.

Horses spend most of the day grazing. They bite grasses and other plants with sharp front teeth. Then they grind the food with large, ridged back teeth. Though a horse's teeth keep growing, their surfaces slowly wear away over many years of use. By looking at a horse's teeth, an expert can usually tell the animal's age.

Ears laid back, a gray mustang mare fights off a stallion. Stallions often nip and chase mares to keep their bands moving and grazing together.

Horse

In many parts of the world, bands of feral (say FEAR-ul) horses roam free. They are the descendants of tame horses. In western North America, feral horses called mustangs descended from animals brought by explorers and settlers as long as 400 years ago.

Wild horses travel in bands of three to twenty animals. A male—called a stallion—guards the females—called mares—and the young foals from such enemies as mountain lions. As his band drinks

Clydesdale: 65 in (165 cm) tall at the withers

◁ *Long leg hairs fluttering, a Clydesdale prances in a corral. Harnessed in teams, Clydesdales and other draft horses once commonly pulled heavy wagons in cities.*

As a horse moves faster and faster, it changes its gait—or the order in which its feet hit the ground. A horse walks at about 4 miles (6 km) an hour. The hoofbeats are slow and even. Some people think that they sound like the words "knick-er-bock-er, knick-er-bock-er." At a trot, a horse moves at about 9 miles (14 km) an hour. The hoofbeats

W A L K

C A N

G A L

Peppy and alert, a Morgan horse, the first American ▷ breed, waits with ears erect. All Morgan horses trace their ancestry to a stallion named for its owner, Justin Morgan.

and grazes, a stallion stands alert. A horse's eyes are placed near the sides of its head. It can see almost in a circle without turning its head. A horse also can turn its ears separately. The position of its ears often shows its mood. When a horse flicks one ear forward and the other ear back, it is alert. Both ears pinned back usually means that the animal is angry.

A stallion keeps his band grazing and moving together. He chases off rival stallions that may try to steal the mares. Stallions *(Continued on page 276)*

Morgan horse: 61 in (155 cm) tall at the withers

sound like "pop-corn, pop-corn." At a canter, a horse travels at about 12 miles (19 km) an hour. The hoofbeats sound like "ap-ple pie (pause) ap-ple pie." The gallop is the fastest gait. Racehorses can reach 42 miles (68 km) an hour. The sounds of the hoofbeats blur together.

T R O T

T   E   R

L   O   P

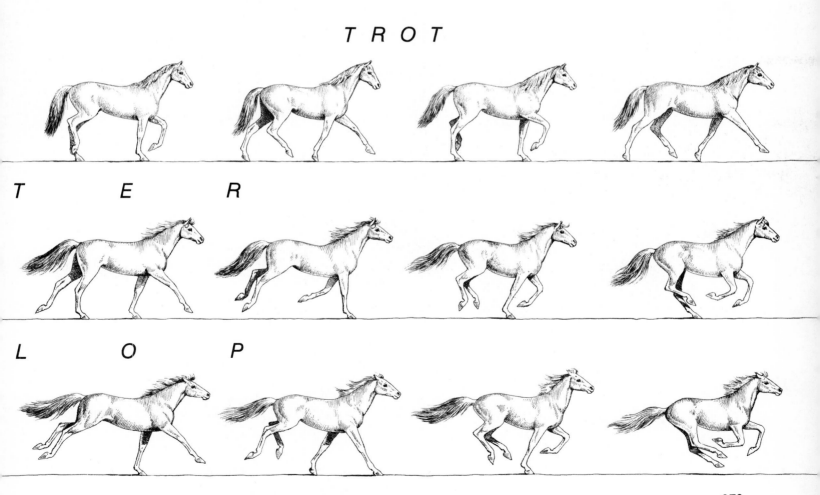

Camargue horse: 56 in (142 cm) tall at the withers

△ Egret stands motionless on the back of a pregnant Camargue mare. These small, white horses have roamed the marshy areas of southern France since Roman times. Swift and agile, they move easily through the marshlands.

Nipping and squealing, Chincoteague pony foals play at ▷ fighting. The ponies belong to a band that lives on a lonely, windblown island off the coast of Maryland and Virginia. Each year, people round up some of the ponies. The animals swim across the narrow bay and are sold on the mainland.

Chincoteague pony: 53 in (135 cm) tall at the withers

Arabian horse: 59 in (150 cm) tall at the withers

△ *Arabian horses race across a field in Sweden. The slender, elegant appearance of these horses has made the breed popular for riding the world over.*

◁ *Stiff manes identify a Przewalski's mare and her foal. Unlike mustangs, Przewalski's horses have no tame ancestors. The stocky animals once roamed central Asia. Today, the few remaining horses live in zoos.*

Przewalski's horse: 50 in (127 cm) tall at the withers

challenge each other by arching their necks and tucking in their chins. They rear, biting and kicking. Finally, one stallion gives up and gallops away.

In the spring, mares give birth—usually to a single foal. The newborn, wobbly on its spindly legs, is up and nursing in an hour.

When the young males, or colts, are about two years old, the stallion drives them away. For a few years, they roam with other young males. Then they are ready to gather bands of their own.

Unlike feral horses, Przewalski's (say per-zhih-VAHL-skeez) horses have no tame ancestors. Named for the Russian explorer who was the first Westerner to study them, these stocky, short-legged horses have stiff manes and a black stripe down their backs. Przewalski's horses once roamed central Asia, but none have been seen in the wild since 1968. Today small groups live protected in zoos.

Most of the horses in the world are domesticated, or tamed. There are now about two hundred breeds of horses. The animals may vary in size, in

HORSE

HEIGHT: 30-69 in (76-175 cm) at the withers—the ridge between the shoulders

WEIGHT: 120-2,200 lb (54-998 kg)

HABITAT AND RANGE: grassy areas in parts of North and South America, Europe, and Australia; domestic horses live in many parts of the world

FOOD: grasses and other plants, including fallen fruit

LIFE SPAN: 25 to 35 years in captivity

REPRODUCTION: usually 1 young after a pregnancy of about 11 months

ORDER: perissodactyls

Iceland pony: 51 in (130 cm) tall at the withers

△ *Iceland pony mares groom one another. They nibble at each other's summer coats with their teeth. In winter, shaggy hair covers these hardy, short-legged horses. The coats protect them in the cold, harsh climate of their island home.*

◁ *White muzzle and eye patches highlight the dark coat of an Arkansas mule, the long-eared offspring of a female horse and a male ass. Mules are tougher and more surefooted than most horses.*

Mule: 58 in (147 cm) tall at the withers

276

color, or in the kinds of jobs they can do. Draft horses such as Belgians and Clydesdales are the largest and strongest of all horses. They have muscular bodies and thick, short legs.

Arabian horses are smaller animals with slender necks, high-set tails, and silky coats. Thoroughbreds, another breed, are famous for their speed at racing. Quarter Horses, which can stop and turn quickly, are used as working horses on ranches. Ponies are the smallest horses. They often have broad chests and strong, sturdy bodies.

Mules are the offspring of a female horse and a male ass. Tough and surefooted, mules often are used for farm work.

Hutia

(*say* OO-TEE-uh)

Bahamian hutia: 12 in (30 cm) long; tail, 3 in (8 cm)

Bahamian hutia rests on a rocky ledge. This kind of hutia survives on only one island in the Bahamas.

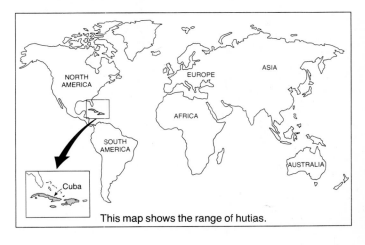

This map shows the range of hutias.

HUTIA

LENGTH OF HEAD AND BODY: **10-20 in (25-51 cm); tail, 1-12 in (3-30 cm)**

WEIGHT: **2-15 lb (1-7 kg)**

HABITAT AND RANGE: **remote forests and rocky regions in parts of the West Indies**

FOOD: **plants and a few small animals such as lizards**

LIFE SPAN: **about 12 years in captivity**

REPRODUCTION: **1 to 3 young after a pregnancy of 2 to 4 months**

ORDER: **rodents**

THOUGH IT ONCE WAS COMMON in the West Indies, the hutia has been heavily hunted by animals and by people. For centuries, the hutia had few enemies, except for the natives who killed it for food. Later, Europeans began to arrive and to bring with them animals that preyed on the hutia. Today the large rodent has disappeared from all but the most rugged parts of the islands.

Hutias look a little like rats, with broad heads, small eyes and ears, and grayish or brownish hair.

Hutias on one island usually are different from those on another island. The Bahamian hutia, for example, has short hair and a short tail. It feeds at night on leaves, bark, and twigs. During the day, it usually sleeps in caves or in small openings between rocks. It survives on only one island in the Bahamas.

Cuban hutias feed during the day. They eat fruit, leaves, and bark as well as such small animals as lizards. Cuban hutias are covered with coarse hair. Although they live in holes in the ground, they often climb trees and sun themselves on the branches. Some Cuban hutias can even wrap their tails around branches.

Hutias have one to three young after a pregnancy of two to four months. The offspring are born with hair and with their eyes open.

Hyena

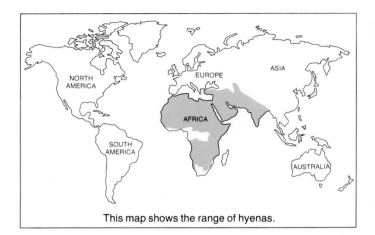

This map shows the range of hyenas.

Two spotted hyenas drink at a water hole. ▷
Spotted hyenas live on the plains of Africa.
They can go for several days without drinking.

▽ *Female spotted hyena takes an afternoon nap (below), as one of her offspring snuggles close and another peeks from their den. Spotted hyenas usually give birth to two cubs in a hole in the ground. For the first few weeks, the female stays close by her cubs. If her young wander away from the den, she carries them back by the napes of their necks (lower right, top). Furry and black at birth, the cubs grow lighter coats after about ten weeks. Spots begin to appear later. Hyena cubs often play (lower right, bottom). They chase each other and splash in water holes. They stay with their mother more than a year.*

AFTER DARK, eerie giggles, yells, and growls drift across the Serengeti Plain of Africa. Spotted hyenas have killed a wildebeest. As they feed, they squabble among themselves over the meat. The biggest hyenas take the best parts of the kill. Younger and weaker animals get the scraps.

A spotted hyena looks like a very large dog. It has a blunt muzzle, rounded ears, and long front legs. The yellow-gray fur on its sloping back is covered with dark spots. But hyenas are not part of the dog family. Their closest relatives are aardwolves.

Until recently, people thought hyenas mainly scavenged for food, that is, they fed on any dead animals they could find. Now scientists have discovered that spotted hyenas are skillful hunters.

As many as eighty hyenas may live together in a group called a clan. The animals often hunt in

Spotted hyena: 45 in (114 cm) long; tail, 13 in (33 cm)

smaller packs—usually ten to thirty hyenas. Keen senses of hearing and eyesight help them track animals at night. Spotted hyenas can run for miles without getting tired. They usually catch what they are chasing. Muscular shoulders and bodies allow them to overpower their prey easily.

Spotted hyenas hunt different animals in different ways. When a pack goes after wildebeests, one hyena will charge into a herd and startle the animals. As the herd runs off, the hyenas watch for a slow or sick wildebeest. Then they all join the chase

In a tug-of-war, spotted hyenas and a lioness fight for ▷
*the meat of a wildebeest that the hyenas killed. When the
lioness heard the giggling calls of the feeding hyenas, she
came to help herself to an easy meal. Several lions can
chase hyenas away from their kill. But a group of hyenas
can usually keep a single lion away.*

▽ *Warming itself in the afternoon sun, a young hyena
in Kenya looks as friendly as a puppy. But hyenas have
jaws and teeth that can shred meat and splinter bones.*

to bring down the victim. When hunting alone, a spotted hyena looks for smaller prey such as a gazelle. It may circle a lake and then wade in to try to attack birds in the water. Or it may stand in the shallows and snap at fish.

Once the kill is made, and after a pack of hyenas begins to feed, the animals make their laughing sounds. Because of these noises, they are often called laughing hyenas. From the sounds, other hyenas know that the hunting pack has found food.

Spotted hyenas can eat almost all of their prey.

Their teeth and powerful jaws can even crush large bones. Their stomach juices can digest bone as well as skin. Hyenas come to some villages in Ethiopia at night and feed on garbage. After the hyenas have eaten, almost nothing is left.

Spotted hyenas often play in pools of water. They may wallow in mud. Sometimes hyenas even store food by dropping it in shallow water. Later, when they are hungry, the animals poke their heads underwater until they find the food.

Each clan lives and hunts in its own territory, or area. Members of a clan mark their territory by scratching the ground and leaving waste. They mark the grass with a smelly substance from glands under their tails. The scent marks warn hyenas from other clans that the territory is occupied. Groups of spotted hyenas may also patrol the borders of the territory. If a hyena from another clan does try to invade, the patrol chases it away. At times two clans of hyenas may meet and battle over a kill.

Clan members often meet at a large den near the middle of the territory. The den is made up of many

281

Brown hyena: 34 in (86 cm) long; tail, 10 in (25 cm)

△ *Shaggy-haired brown hyena lopes across scrubby dry plains in southern Africa. Brown hyenas usually travel alone. They may go long distances in search of such food as small mammals, fruit, or dead animals.*

burrows, connected by tunnels. At the den, hyenas often greet each other by sniffing like dogs. In this way they recognize members of their clan. Hyenas have many ways of communicating. One animal may walk on its wrists in front of another hyena that is more important. Or a hyena may raise its whisk-broom tail to show excitement. Hyenas also use other calls besides laughter. Loud grunts often signal a threat.

Female hyenas give birth to twin cubs in shallow holes a short distance from the clan den. About two weeks later, a female hyena carries her cubs to the den. There all the cubs live together. The young stay with their mothers and nurse for more than a year. After a few months, however, they follow along on hunts and snatch scraps.

Another kind of hyena, the striped hyena, roams parts of Africa and Asia. Usually striped hyenas stay by themselves, but they may live together when they are raising young. During the day, they rest in holes in the ground. At night, they wander great distances in search of food. The striped hyena is mainly a scavenger of dead animals. But it may hunt for small mammals. It also eats fruit and

insects. Unlike the spotted hyena, the striped hyena does not laugh or whoop.

The rare brown hyena of southern Africa also lives by itself. The brown hyena is smaller than its spotted relative, but it has longer hair. Dark stripes circle its legs. It roams widely looking for live or dead animals. Sometimes it feeds on melons and ostrich eggs. The brown hyena is seldom seen by people. Scientists are trying to learn more about its behavior in the wild.

HYENA

LENGTH OF HEAD AND BODY: 34-59 in (86-150 cm); tail, 10-14 in (25-36 cm)

WEIGHT: 82-190 lb (37-86 kg)

HABITAT AND RANGE: dry plains and brushy areas in parts of Africa, the Middle East, and Asia

FOOD: mammals, fruit, and insects

LIFE SPAN: as long as 25 years in the wild

REPRODUCTION: usually 2 young after a pregnancy of about 3 or 4 months

ORDER: carnivores

▽ *Striped hyena stretches its long, thick neck. Smaller than its spotted relative, this hyena has the typical powerful shoulders and sloping back of the hyena family. Striped hyenas live in parts of Africa and Asia.*

Striped hyena: 40 in (102 cm) long; tail, 11 in (28 cm)

Hyrax

Rock hyrax: 19 in (48 cm) long

Young rock hyraxes, watched over by an adult, huddle together on a boulder. Hyraxes often cooperate in caring for young. One adult tends the offspring while other adults feed.

THEY DON'T LOOK MUCH ALIKE, but the tiny hyrax and the huge elephant *are* related. Scientists believe that both come from a common ancestor that lived about 55 million years ago. One glance at their feet shows that they are relatives. On their toes, the elephant and the hyrax both have flat nails that are almost like hooves. Some African tribes call hyraxes "little brothers of elephants."

The hyrax looks like a guinea pig. But the hyrax is not a rodent. It is so different from all other animals that scientists have placed it in an order, or group, all by itself.

There are three kinds of hyraxes. The rabbit-size rock hyrax is the largest. The tree hyrax and the bush hyrax are smaller.

Colonies of rock hyraxes and bush hyraxes live in rocky areas in parts of Africa and the Middle East. Sometimes rock hyraxes and bush hyraxes live side by side. During the day, the animals search for food. Some scientists think that the animals can live together because they usually do not eat the same foods. Rock hyraxes nibble mostly grasses. Bush hyraxes climb into trees and bushes and feed on leaves. While they eat, one of the hyraxes watches for enemies—leopards, eagles, and snakes.

Tree hyraxes live in the forests of Africa. These shy little animals become active at night, feeding on leaves and buds. They climb easily. Like the other hyraxes, tree hyraxes have rubbery pads on the soles of their feet that give them a good grip.

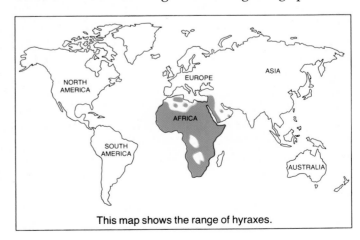

This map shows the range of hyraxes.

283

Bush hyrax: 17 in (43 cm) long

Tree hyrax: 17 in (43 cm) long

△ *Sound the alarm! A rock hyrax warns others of danger. If an enemy such as an eagle, a snake, or a leopard comes near the colony, the hyrax will give a shrill cry.*

Little is known about the habits of the tree hyrax. But many people in Africa have heard the animal's call. Just after dark, a tree hyrax begins to make a soft, ringing sound. It repeats this call several times, getting louder with each cry. Finally, the hyrax gives a piercing scream. Then there is silence until other hyraxes begin their calls. Their cries are heard throughout the night. Scientists think that the male tree hyrax uses its call to find a mate or to say, "This part of the forest belongs to me."

Hyraxes usually have litters of one to four young after a pregnancy of about seven and a half months. The young can run and jump soon after birth. They nurse for as long as six months. Rock hyraxes and bush hyraxes sometimes eat grasses or leaves when they are only two days old.

△ *Termite mound offers a spot for several bush hyraxes to sun themselves. If danger threatens, they will dart into the holes in the mound and hide.*

◁ *Rock hyrax (far left) nibbles on a plant. Though these animals prefer grass, they will eat fruit, flowers, buds, and leaves, too. During the dry season, rock hyraxes use their curving front teeth to strip bark from bushes and small trees. Young tree hyraxes (left) in Africa peer out from among leaves. Hyraxes nurse for as long as six months. But some can eat grasses and leaves soon after birth. Young hyraxes scamper, jump, and climb as easily as adults do. Rubbery pads on the soles of their feet help them cling to rocks and branches.*

HYRAX

LENGTH OF HEAD AND BODY: **16-22 in (41-56 cm)**

WEIGHT: **4-11 lb (2-5 kg)**

HABITAT AND RANGE: **rocky areas and forests in parts of Africa and the Middle East**

FOOD: **grasses, leaves, buds, stems, bark, and fruit**

LIFE SPAN: **as long as 12 years in captivity**

REPRODUCTION: **1 to 4 young after a pregnancy of about 7½ months**

ORDER: **hyracoids**

I

Ibex

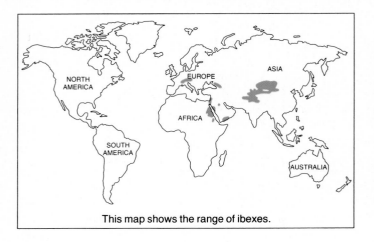

This map shows the range of ibexes.

IBEX

HEIGHT: 26-41 in (66-104 cm) at the shoulder

WEIGHT: 70-275 lb (32-125 kg)

HABITAT AND RANGE: mountainous and rocky regions of Europe, Asia, and northeastern Africa

FOOD: grasses, herbs, shrubs, and other plants

LIFE SPAN: as long as 22 years in captivity

REPRODUCTION: 1 or 2 young after a pregnancy of 5 or 6 months, depending on species

ORDER: artiodactyls

HIGH IN THE MOUNTAINS, the ibex climbs among rocky crags and cliffs. It grips the slopes with the soles of its split hooves. A kind of wild goat, the surefooted ibex seldom slips or falls—even though it carries a heavy weight on its head.

The male ibex has a pair of ridged horns that curve as much as 5 feet (152 cm) into the air. The horns can weigh 20 pounds (9 kg) The female has horns too, but they are shorter and lighter. The horns of an ibex keep growing year after year. Each year, a new ridged section is added. The older an ibex is, the longer its horns are. People can count the sections and find out how old an ibex is.

Ibexes have short, brownish coats. The males have beards. Except for the horns and beards, ibexes look much like their relatives, bighorn sheep. Ibexes live in mountainous and rocky regions of Europe, Asia, and northeastern Africa.

One kind of ibex lives high on treeless mountain slopes of the Alps in Europe. In winter, these Alpine ibexes often come partway down the mountainsides to avoid the deep snows higher up. The animals seek out slopes facing south. It is sunnier and warmer there than on other slopes. Ibexes also try to find places where the wind blows the snow off grasses and other plants.

Alpine ibexes graze on grass when they can find it. But they also eat herbs, shrubs, and other plants. In the spring, before they return to higher ground, the ibexes sometimes go down the slopes to areas where trees grow. There they feed on green shoots.

Ibexes normally spend about half of each day filling their stomachs with partially chewed plants. Later, while they rest, they bring up wads of this food—called cuds—from their stomachs. They chew the cuds thoroughly and swallow the food again so that it can be further digested. Some other animals that chew cuds are antelopes, cows, goats, and sheep. Read about these animals under their own headings. When it is very windy, ibexes sometimes go to sheltered spots to spend the night.

Except during the mating season in late fall and early winter, adult male and female Alpine ibexes usually live apart from each other. Most males roam together in groups. Sometimes they gather in herds of as many as forty to fifty animals. Old males often remain alone. Females and their kids form herds of about a dozen animals.

The herds of males break up during the mating season. The biggest and strongest male ibexes mate with more females. Smaller and weaker males rarely challenge the larger ones. But males of the same size sometimes fight to find out which is the stronger. Two male ibexes cross their massive horns and push against one another. They rise on their hind legs and butt their heads. The crash of their horns can sound like rocks banging together. The clash can be fierce

High on a snow-covered ridge in Italy's Gran Paradiso National Park, male Alpine ibexes graze. The animals use their hooves to uncover grass beneath the snow.

Alpine ibex: 34 in (86 cm) tall at the shoulder

Ibex

enough to break a horn. In May or June, about five months after mating, female Alpine ibexes give birth to young. Usually only one kid is born, but sometimes a female ibex has twins. After giving birth, a mother will lick her kid. She becomes familiar with its smell, so she will be able to identify her kid in the herd.

Newborn ibexes weigh about 8 pounds (4 kg). They develop quickly and can walk by the end of their first day. Soon after, they can run and jump. Within a few weeks, young ibexes can balance on the edges of rocky cliffs. They also eat small pieces of plants. The frisky kids take turns pushing each other off rocks. Sometimes they even playfully jump on their mothers' backs.

Female ibexes keep watch over their young. If a mother loses sight of her kid, she may "baaa" like a sheep. Her kid answers with a "baaa" of its own. When she finds her young, the female may sniff it to make sure it is hers.

In the Alps, golden eagles still prey on young ibexes. In other areas, leopards, bears, and wolves

Lone male ibex watches from a mountainside in Italy. ▷
Old males often do not roam with a herd. They travel by themselves, except at mating time.

▽ *Female Alpine ibex and two kids graze on a slope in the Gran Paradiso park. Alpine ibexes usually give birth to one young at the end of May or in June.*

Male Nubian ibexes pause on a rocky trail. Knobby, ridged sections on their massive horns show their age.

Nubian ibex: 33 in (84 cm) tall at the shoulder

hunt these horned animals. People too have hunted Alpine ibexes. Though the animal once was common, by the early 1800s it was almost extinct.

One reason the Alpine ibex became so rare is that some people believed that parts of the animal could cure various diseases. Balls of undigested hair from an ibex's stomach, for example, were thought to protect people from cancer.

In 1854, King Victor Emmanuel II began to protect the few remaining Alpine ibexes. He set up a preserve for the ibexes in a part of the Italian Alps known as the Gran Paradiso. This preserve later became a national park. Ibexes survived and multiplied. Animals from the Gran Paradiso were sent to the mountains of other European countries. Today thousands of ibexes roam the Alps again.

Ibex

Nubian ibexes carefully pick their way along the side ▷
of a steep cliff on a preserve in Israel. Much of the land
these animals once roamed has been taken over for
grazing by domestic sheep and goats.

▽ Young Nubian ibexes (below) push and shove as
they play among rocks. Ibex kids may play much of the
time. Adult ibexes spend the day feeding and resting. An
adult male (bottom) points his face toward the sun and
leans back on the tips of his horns. While resting, ibexes
often chew cuds—partly digested wads of food brought
up from their stomachs.

The Alpine ibex is one of several kinds of ibexes.
All ibexes look and behave alike, but there are a few
differences. For example, the Nubian ibex is smaller
than the Alpine ibex, and its horns are longer and
thinner. The Nubian ibex lives in rocky desert regions around the Red Sea, in other parts of Egypt, in
Jordan, and in other countries. In Israel, the Nubian
ibex also lives protected on preserves.

Walia ibexes live in Ethiopia. They are larger

than Nubian ibexes. Walia ibexes have suffered as their range has been turned into farmland. Domestic sheep and goats also have taken over more and more of the land where these ibexes feed. Now the Walias are close to extinction.

The Siberian ibex, largest of all ibexes, is found in the mountains of central Asia. It can live as high as 16,000 feet (4,877 m). This animal is well adapted, or suited, to the cold weather of its home. Female Sibe-

rian ibexes bear their young after six months—almost a month longer than other ibexes. This means Siberian ibex kids are born in June or July, when the weather is warm enough for them to survive.

In ancient times, ibexes were often kept by kings in Persia and in Egypt. Small statues of ibexes have been found in the tombs of Persian royalty. And images of ibexes have even been found in the tomb of King Tutankhamun.

Impala

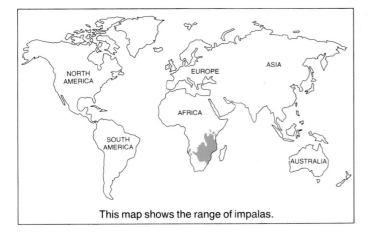

This map shows the range of impalas.

SENSING A LION NEARBY, an impala barks a sharp alarm that warns the herd. In a flash, fifty antelopes scatter in several directions across a plain in Africa. The startled and confused enemy hesitates. In that instant, the graceful, reddish brown animals disappear into the cover of shrubs and bushes.

When frightened, impalas spring into action. In a series of racing strides and long-distance jumps, they dash for safety. Soaring upward as high as 10 feet (305 cm), they clear tall grass and bushes—and even jump over other impalas. Their long, slender legs and muscular thighs enable them to travel as far as 30 feet (9 m) in a single leap.

Impalas are medium-size antelopes. They measure about 3 feet (91 cm) tall at the shoulder. Lightly built for speed and jumping, they weigh about 125 pounds (57 kg). Male impalas have sharp, curving horns that sweep backward and then turn upward. Females have no horns. Learn about other kinds of antelopes on page 52.

Impalas range over parts of eastern, central, and southern Africa. They live on grasslands, in brushy areas, and in open woodlands. Throughout the rainy season, impalas in some regions may graze in large herds of hundreds of animals. In the dry season, they must browse, or nibble on leaves and shoots of bushes. With food in short supply, the

IMPALA

HEIGHT: 33-39 in (84-99 cm) at the shoulder

WEIGHT: 88-165 lb (40-75 kg)

HABITAT AND RANGE: grasslands, shrubby areas, and open woodlands of eastern, central, and southern Africa

FOOD: grasses, herbs, bushes, and shrubs

LIFE SPAN: as long as 17 years in captivity

REPRODUCTION: usually 1 young after a pregnancy of about 7 months

ORDER: artiodactyls

◁ *Trade-off! Birds known as oxpeckers find a comfortable roost and also keep an impala free of ticks and insects. The young male appears undisturbed by his passengers. Face-to-face, two male impalas (far left) meet before a battle that will test their strength. Males of the same age and size may fight frequently, pushing at each other with their horns.*

herd usually breaks up into smaller groups. The animals roam in search of water holes. At other times of the year, water is usually available nearby.

In a few places, impalas mate at any time of the year. Older males mark out territories, or areas, and defend them from rivals. When females wander into a male's territory, he rounds them up. By herding them with a honking sound, he may keep them for days before they move on.

Adult males that have no territories often join younger males in groups known as bachelor herds. Bachelors frequently fight one another to find out which is stronger. The strongest bachelors then challenge the males with territories.

A battle for territory usually begins slowly. At first, the rivals parade about, horns held high. Sometimes a male will be able to force a challenger

Exploding into action, impalas leap away from danger. ▷
Swiftly, the herd flees across a grassy plain in Tanzania.
An impala will jump over almost any object in its path—
including another impala.

▽ *Looking for mates, a male impala uses loud honking*
noises to round up females. Despite his efforts, the females
may soon leave his territory for that of a neighbor.

out of his territory without a real fight. In more serious battles, the males meet head to head. They lunge forward and try to jab each other with their sharp horns. They also lock horns, pushing and twisting their heads. But they seldom get badly hurt. The layer of skin covering a male's head and neck is especially thick and protects him. When one of the males can no longer keep up the struggle, he runs away. The winner may pursue him. Eventually the loser returns to the bachelor herd.

Holding a territory can be tiring for a male impala. He must fight off intruders, chase bachelors, herd the females, and mate with them. During this time, he may eat little.

About seven months after mating, a female leaves the herd for a secluded place. There she gives birth—usually to one young. After a few days, the mother and her young join a herd of females and offspring. A young female will remain with the females. A young male will join a bachelor herd.

Dark stripes on its rump flash as a male impala springs gracefully into the air. Strong, slender legs equip the animal for long, high leaps.

295

J

Jackal

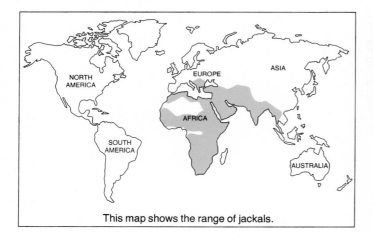

This map shows the range of jackals.

Golden jackal: 32 in (81 cm) long; tail, 10 in (25 cm)

△ *Keeping watch over a pup, golden jackals groom each other. Young sometimes roll over so parents can groom their bellies.*

SOON AFTER SUNSET, the howls of golden jackals echo across the plains of northern Africa. Jackals hunt mainly at night. With loud calls, these members of the dog family signal the beginning of their hunt. At sunrise, after a night of hunting, jackals again begin their eerie howls. Jackals communicate with many other kinds of sounds. They bark softly to warn that lions are nearby. Pups whine when they greet their parents.

Golden jackals roam the grassy plains of Africa and the brushy woodlands of India. Other kinds of jackals are found only in Africa. The silver-backed, or black-backed, jackal makes its home in brushy woodlands. The tan-and-red fur of this bold animal is topped with a saddle of black-and-white hairs on its back. The side-striped jackal often lives in brushy areas where the animal can find cover more easily. But it may hunt on the plains.

Jackals usually hunt alone for small mammals and insects. When they pair up or hunt in small groups, they can bring down bigger game, such as a young gazelle. Jackals often feed on the remains of dead animals or on kills that have been made by

◁ *Fur bristling and ears laid back, a golden jackal stands its ground against a Cape hunting dog. The jackal tried to steal the wild dog's kill. Jackals often feed on the prey of other meat eaters.*

Glassy surface of a water hole ▷ *mirrors a silver-backed, or black-backed, jackal as it sniffs the ground. The animal can find out about recent visitors to the area by scent. Its keen senses help this hunter find food.*

Silver-backed jackal: 28 in (71 cm) long; tail, 14 in (36 cm)

Jackal

△ Sitting up, two nine-week-old silver-backed jackals nurse on their mother's milk. The female watches the grasslands as her pups drink. She will fiercely defend them against such enemies as hyenas and eagles.

Flying leap after a rat takes a silver-backed jackal ▷ high above the grass.

▽ Ready for solid food, a young silver-backed jackal carries a rat killed by the pup's mother. Like most wild dogs, jackals feed their young by swallowing food and taking it back to the den in their stomachs. There they bring it up for the pups to eat.

other meat eaters. After a lion has fed from a kill, jackals gather round to eat the scraps.

A pair of male and female jackals stays together for years, or even for life. Before giving birth to her young, the female takes over a den, often an abandoned aardvark burrow. One to eight helpless pups are born in a litter. Both parents help care for the young. At two weeks of age, the pups begin to explore their world. At three months of age, they tag along on hunting trips. Young silver-backed jackals and golden jackals often stay with their parents for a year and help raise the next litter.

JACKAL

LENGTH OF HEAD AND BODY: 26-42 in (66-107 cm); tail, 8-16 in (20-41 cm)

WEIGHT: 14-33 lb (6-15 kg)

HABITAT AND RANGE: plains, brushy woodlands, and deserts in parts of Africa and Asia

FOOD: rodents, small antelopes, birds, reptiles, insects, fruit, berries, grasses, and remains of dead animals

LIFE SPAN: as long as 16 years in captivity, depending on species

REPRODUCTION: 1 to 8 young after a pregnancy of about 2 months

ORDER: carnivores

Jackrabbit

Jackrabbits are hares. Read about them on page 250.

Jaguar

(*say* JAG-wahr)

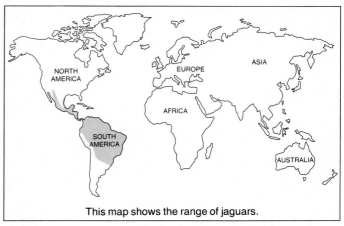

This map shows the range of jaguars.

FOR HUNDREDS OF YEARS, the mysterious jaguar has played an important role in the religions of Indians from Mexico into South America. For some, the Jaguar God of the Night was a ruler of the underworld. Its spotted coat represented the stars in the night sky. In spite of all the stories, though, scientists still know little about this big cat.

The jaguar is the largest cat in the Western Hemisphere. It prowls rain forests, marshes, dry scrublands, and grasslands from Mexico into Argentina. An expert swimmer, the jaguar wades into water to catch a fish or to chase a caiman—a small relative of the alligator. The jaguar also stalks prey in high grass or bushes. There it catches such animals as capybaras, peccaries, deer, and tapirs. The cat creeps close to its prey and then swiftly pounces. It seizes the animal with muscular forelegs and kills it with a bite in the neck or head. The jaguar is extremely strong. It may drag even heavy prey to a sheltered spot some distance away.

The jaguar has a broad head, very powerful jaws, large shoulders, and sturdy legs. Its short, stiff fur is usually golden or reddish orange. Rings of small black dots pattern its coat. Because these rings

Powerful muscles rippling under brilliantly patterned fur, a jaguar guards its realm. Even a faint rustle of leaves or a snapped twig puts the cat on the alert.

301

Jaguars sun themselves high on a tree limb. From their lofty perch, these skilled climbers scan the ground below. In the wild, jaguars have little contact with one another. These cats, however, share their grounds with other jaguars in a zoo in Brazil. No one knows the number of jaguars left in the wild. Hunters kill many of the cats each year for their richly spotted fur.

are rose-shaped, they are called rosettes. A jaguar's rosettes often have dark spots in the centers.

Jaguars usually weigh between 100 and 250 pounds (45-113 kg) and measure about 6 feet (183 cm) from head to rump. On the average, they are smaller than lions and tigers but larger than leopards. Scientists group all four kinds of animals together as big cats. Find out about leopards, lions, and tigers under their own headings.

All four big cats can roar, but their roars do not sound alike. The jaguar's roar resembles a loud cough, repeated several times. Jaguar hunters in Brazil imitate this sound by grunting into a hollow gourd. The jaguar is attracted to the sound.

A jaguar lives alone in a home range—an area it marks out for itself. Home ranges are often large, covering many square miles. Although the ranges of two or more jaguars may overlap, the cats rarely meet. They signal their whereabouts by scratching trees and leaving waste.

Male and female jaguars meet at mating time. About three months after mating, a female bears her cubs—usually two or three in a litter. The young jaguars are born blind and helpless. For more than a year, they stay and hunt with their mother. Then, gradually, they spend more and more time on their own. Finally, they leave their mother's home range and find one of their own. At about three years of age, jaguars are fully grown.

No one knows exactly how many of these cats still roam in the wild. Although jaguars once lived in the southern United States, they were hunted to extinction there. Farmers and ranchers thought the cats were a threat to livestock. Today ranchers in South America still kill jaguars to protect their cattle. The cats also are hunted for their fur.

JAGUAR

LENGTH OF HEAD AND BODY: **5-6 ft (152-183 cm); tail, 20-31 in (51-79 cm)**

WEIGHT: **100-250 lb (45-113 kg)**

HABITAT AND RANGE: **rain forests, marshes, scrublands, and grasslands from Mexico into Argentina**

FOOD: **rodents, peccaries, deer, tapirs, fish, and cattle**

LIFE SPAN: **as long as 20 years in captivity**

REPRODUCTION: **1 to 4 young after a pregnancy of 3 or 4 months**

ORDER: **carnivores**

Jerboa

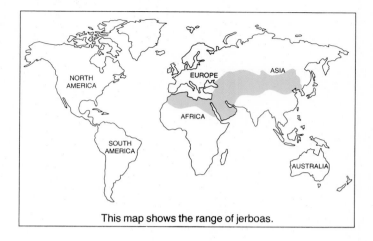

This map shows the range of jerboas.

JERBOA

LENGTH OF HEAD AND BODY: 2-6 in (5-15 cm); tail, 3-10 in (8-25 cm)

WEIGHT: 2-4 oz (57-113 g)

HABITAT AND RANGE: deserts and dry plains in parts of Asia and Africa

FOOD: plants, seeds, and insects

LIFE SPAN: about 6 years in captivity

REPRODUCTION: 2 to 6 young after a pregnancy of about a month

ORDER: rodents

Standing on its long hind legs, a desert jerboa listens for danger. Its ear may have been nicked in a fight with another jerboa. Usually the animal escapes its enemies by leaping away. When jumping, the jerboa uses its long tail for balance.

Desert jerboa: 6 in (15 cm) long; tail, 10 in (25 cm)

LOOKING LIKE A TINY KANGAROO, the sand-colored jerboa springs across bare ground. Using its strong hind legs and feet to push off, this rodent can leap as far as 10 feet (305 cm). Such long-distance jumps are amazing for an animal that may measure only 6 inches (15 cm) long!

There are about 25 species, or kinds, of jerboas. They live in deserts and on dry plains in parts of Asia and Africa. Jerboas are active at night. They look for plants and seeds. Some jerboas also feed on beetles and other insects.

With their short front legs and their strong teeth, jerboas dig burrows in the ground. These underground homes protect them from both hot and cold weather. The burrows also help keep the animals safe from enemies such as owls and foxes. From inside their burrows, jerboas sometimes block the entrances with sand or soil. Using their snouts, they shove loose dirt into the openings.

Some jerboas curl up in their burrows during the winter. There they hibernate (say HYE-bur-nate), or sleep. Their heart rates drop, and their breathing slows down. Some jerboas escape the heat of summer by sleeping this way. Summer sleep is called aestivation (say es-tuh-VAY-shun).

Females give birth to two or three litters a year. Each litter contains two to six young.

Young chimpanzee in Africa balances between two vines.

Photographers' Credits

COVER Irven DeVore/Anthro-Photo. **(1)** Keith Gunnar. **(2-3)** Thomas Nebbia. **(3)** PITCH/P. Montoya. **(4)** Kenneth W. Fink/ARDEA LONDON. **(6-7)** Steven C. Kaufman. **(10-11)** Robert Caputo. **(12)** *left,* Loren McIntyre; *top right,* Karl Weidmann; *bottom right,* Hans & Judy Beste. **(13)** R. S. Virdee/Grant Heilman Photography. **(14)** Jeff Foott. **(15)** *top,* Merlin D. Tuttle; *bottom left,* Bannister/NHPA; *bottom right,* Hans & Judy Beste. **(16)** *top,* Stanley Breeden; *bottom,* Stephen J. Krasemann/DRK Photo. **(18)** *top,* David Hiser; *bottom left,* Larry R. Ditto; *bottom right,* Robert P. Carr. **(19)** ANIMALS ANIMALS/Stouffer Productions. **(28)** *top,* Hans & Judy Beste; *bottom,* Joan Root. **(29)** *top,* © Reinhard Kunkel; *bottom,* Merlin D. Tuttle-J. Scott Altenbach. **(30)** Tupper Ansel Blake. **(31)** Wolfgang Bayer. **(32)** © Jean-Paul Ferrero. **(33)** *top,* Fred Bruemmer; *bottom left,* Tom & Pat Leeson; *bottom right,* © Peter Johnson. **(34)** George W. Frame. **(34-35)** Sven-Olof Lindblad. **(35)** *bottom left,* Wolfgang Bayer; *right,* Stanley Breeden. **(36)** *top,* Thomas D. Mangelsen; *bottom left,* Helen Rhode; *bottom right,* John M. Burnley. **(37)** Rolf O. Peterson. **(38)** *left,* Keith Gunnar; *right,* Jeff Foott. **(39)** *top,* James K. Morgan; *bottom left,* Greg Beaumont; *bottom right,* Jen & Des Bartlett. **(40)** Merlin D. Tuttle. **(41)** *top right,* Rollie Ostermick; *bottom left,* Bates Littlehales/N.G.S. Photographer; *bottom right,* Christopher Springmann. **(42-43)** George D. Lepp. **(43)** *top,* Wolfgang Bayer; *bottom,* George D. Lepp. **(44-45)** Alan Root. **(46)** Alan Root. **(47)** PITCH/Francois Gohier. **(48)** Francois Gohier. **(49)** Loren McIntyre.

(50) Wolfgang Bayer. **(50-51)** Francisco Erize/BRUCE COLEMAN INC. **(51)** *left,* George B. Schaller; *right,* Jen & Des Bartlett. **(52-53)** Clem Haagner. **(54-55)** *top,* © Peter Johnson; *center,* PITCH/Jean-Claude Carton. **(54)** *left,* © Frederic/Jacana/The Image Bank; *right,* Clem Haagner/BRUCE COLEMAN INC. **(55)** *top,* © 1974 Gail Rubin; *bottom.* © Reinhard Kunkel. **(56)** *top and center,* © Peter Johnson; *bottom,* Kenneth W. Fink/ARDEA LONDON. **(57)** Peter Davey/BRUCE COLEMAN LTD. **(58)** *left,* Kenneth W. Fink/BRUCE COLEMAN INC.; *right,* Belinda Wright. **(58-59)** © 1979 Patricia D. Moehlman. **(59)** *top,* PITCH/P. Montoya; *bottom left and right,* Gerald Cubitt. **(60-61)** George Holton/OCELOT INC. **(62)** Laurance B. Aiuppy © 1980. **(63)** *top and bottom,* Laurance B. Aiuppy © 1980; *center,* Larry R. Ditto/BRUCE COLEMAN INC. **(64)** *top,* Francois Gohier; *bottom,* George B. Schaller. **(65)** *top,* N. Smythe/National Audubon Society Collection/PR; *bottom,* Tracy S. Carter. **(66)** © 1980 Patricia D. Moehlman. **(67)** David Cavagnaro. **(68)** Stanley Breeden; *inset,* © Gail Rubin. **(69)** Anthony & Elizabeth Bomford/ARDEA LONDON. **(70-71)** Tom McHugh/National Audubon Society Collection/PR. **(72)** *left,* Charles G. Summers, Jr./BRUCE COLEMAN INC.; *right,* ANIMALS ANIMALS/Ernest Wilkinson. **(73)** Robert P. Carr/BRUCE COLEMAN INC. **(74-75)** PITCH/D. Heuclin. **(75)** *top,* M.P.L. Fogden/BRUCE COLEMAN INC.; *bottom,* C. B. & D. W. Frith/BRUCE COLEMAN INC. **(76)** *left,* Douglas Baglin/NHPA; *right,* Stanley Breeden. **(77)** Merlin D. Tuttle. **(78-79)** Merlin D. Tuttle. **(78)** *top and center,* Merlin D. Tuttle. **(79)** *top and bottom right,* Merlin D. Tuttle; *bottom left,* Hans & Judy Beste. **(81)** J.A.L. Cooke/Oxford Scientific Films. **(82)** *top,* PITCH/Billes; *center left,* Adrian Warren/ARDEA LONDON; *center right,* Hans & Judy Beste; *bottom,* Merlin D. Tuttle. **(83)** *top,* Stanley Breeden; *far left,* Rene-Pierre Bille; *left,* Kenneth W. Fink/BRUCE COLEMAN INC.; *right,* M.P.L. Fogden/BRUCE COLEMAN INC.; *far right,* PITCH/Jean-Paul Ferrero. **(84-85)** Martin W. Grosnick. **(85)** Stephen J. Krasemann/DRK Photo. **(86)** Patrick E. Powell. **(86-87)** Jeff Foott; *inset,* Steven C. Kaufman. **(88)** *top,* Sonja Bullaty-Angelo Lomeo/courtesy Time-Life Books; *bottom,* Douglas H. Chadwick. **(89)** *left,* Jen & Des Bartlett; *right,* Lynn L. Rogers. **(90)** *top,* George W. Calef; *bottom,* Mickey Sexton/AlaskaPhoto. **(91)** © Canada Wide. **(92)** *left,* Jean-Paul Ferrero; *right,* Tom McHugh/National Audubon Society Collection/PR. **(93)** *left,* PITCH/Binois; *top and bottom right,* Bernard Peyton. **(94)** Jen & Des Bartlett. **(95)** *top,* Wolfgang Bayer/BRUCE COLEMAN INC.; *bottom,* Jen & Des Bartlett. **(96)** *left,* Wolfgang Bayer; *right,* Jen & Des Bartlett. **(97)** Wolfgang Bayer. **(98)** *top,* Leonard Lee Rue III; *bottom,* Rollie Ostermick. **(99)** Tom McHugh/National Audubon Society Collection/PR.

(100-101) Laurance B. Aiuppy © 1980. **(101)** John M. Burnley. **(102)** Erwin & Peggy Bauer. **(103)** *top left,* Jeff Foott; *top right,* Jim Brandenburg; *bottom,* Francisco Erize/BRUCE COLEMAN INC. **(104)** Charles G. Summers, Jr. **(105)** *top,* Leonard Lee Rue III; *center,* Gary R. Zahm; *bottom,* L. West. **(106-107)** PITCH/P. Montoya. **(108)** *top,* PITCH/G. Vienne-F. Bel; *bottom,* George W. Frame. **(109)** Dieter & Mary Plage/BRUCE COLEMAN LTD. **(110)** *top,* Jim Brandenburg; *bottom,* Loren McIntyre. **(111)** *top,* Jean-Paul Ferrero; *bottom,* Bob Abrams/BRUCE COLEMAN LTD. **(112)** *top,* Bob Campbell; *bottom,* Gary Milburn/TOM STACK & ASSOCIATES. **(113)** Lee Lyon/BRUCE COLEMAN LTD. **(114-115)** Victor Englebert. **(116)** Gerald Cubitt. **(117)** *top,* Victor Englebert; *bottom,* Leonard Lee Rue III. **(118)** George B. Schaller. **(119)** *top,* Loren McIntyre; *bottom,* Sven-Olof Lindblad. **(120)** George W. Calef. **(121)** Helen Rhode. **(122)** Martin W. Grosnick. **(122-123)** Helen Rhode. **(124-125)** Rollie Ostermick. **(125)** *top,* Steven C. Wilson/ENTHEOS; *bottom,* Farrell Grehan. **(126-127)** Wolfgang Bayer. **(128-129)** Maurice Hornocker. **(129)** *top,* Maurice Hornocker; *bottom,* Charles G. Summers, Jr. **(130)** *top,* Livan & Rogers/BRUCE COLEMAN INC.; *bottom,* A. A. Geertsman. **(131)** *top,* Karl Weidmann; *bottom left,* © Jean-Paul Ferrero; *bottom right,* Francisco Erize. **(132)** *left,* Belinda Wright; *right,* Stanley Breeden. **(133)** *top left and bottom right,* Stanley Breeden; *top right,* Belinda Wright. **(134)** George B. Schaller. **(135)** *left,* Bob Campbell; *right,* Tadaaki Imaizumi/ORION PRESS. **(136)** *top,* Jane Burton/BRUCE COLEMAN INC.; *bottom left,* © Stephen Green-Armytage 1981; *bottom right,* Hans Reinhard/BRUCE COLEMAN LTD. **(137)** © Stephen Green-Armytage 1981. **(138)** © Jean-Paul Ferrero. **(139)** © Jean-Paul Ferrero. **(140-141)** © Reinhard Kunkel. **(141)** Robert Caputo. **(142-143)** Robert Caputo. **(143)** *top,* Norman Myers/BRUCE COLEMAN INC.; *bottom,* George W. Frame. **(144)** *top,* Robert Caputo; *center,* Robin Pellew; *bottom,* Clem Haagner. **(145)** Bob Campbell. **(146)** Derek Bryceson. **(147)** *top,* Derek Bryceson; *bottom,* Jane Goodall. **(148)** *top,* Hugo Van Lawick; *bottom,* Jane Goodall. **(149)** Derek Bryceson; *inset,* Jane Goodall.

(150) *top and bottom left,* Jane Goodall; *top right,* Hugo Van Lawick. **(151)** Russ Kinne, National Audubon Society Collection/PR. **(152)** L. West. **(153)** *top,* Wayne Lankinen/BRUCE COLEMAN LTD.; *bottom,* Leonard Lee Rue III/BRUCE COLEMAN LTD. **(154)** Anthony and Elizabeth Bomford/ARDEA LONDON. **(155)** *top,* Rod Williams/BRUCE COLEMAN LTD.; *bottom,* BRUCE COLEMAN INC. **(156)** *top,* Jeff Foott; *bottom,* Francois Gohier. **(157)** *left,* Stephen J. Krasemann/DRK Photo; *right,* Jen & Des Bartlett. **(158)** Jim Brandenburg/BRUCE COLEMAN INC. **(159)** *top,* Jonathan T. Wright/BRUCE COLEMAN INC.; *bottom left,* Harste/BRUCE COLEMAN INC.; *bottom right,* J. B. Blossom/NHPA. **(160)** Stanley Breeden. **(161)** *top,* Loren McIntyre; *bottom left,* S. C. Bisserot; *bottom right,* Tom McHugh/National Audubon Society Collection/PR. **(162)** Gary M. Banowetz. **(163)** *left,* Glenn D. Chambers; *right,* Jane M. Pascall. **(164)** Jon Farrar. **(164-165)** Charles G. Summers, Jr. **(165)** David Hiser. **(166-167)** Bildarchio Paysan Stuttgart. **(167)** PITCH/J. C. Bacle. **(168)** © Jean-Paul Ferrero. **(169)** *top,* Jack Fields/National Audubon Society Collection/PR; *bottom,* Belinda Wright. **(170-171)** Thase Daniel. **(171)** William J. Weber. **(172)** *top,* Francisco Erize; *left,* George H. Harrison; *right,* Jean-Paul Ferrero. **(173)** *left,* George Holton/OCELOT INC.; *center,* Stanley Breeden; *right,* George H. Harrison. **(174)** Charles G. Summers, Jr. **(174-175)** Belinda Wright. **(175)** *top,* Francois Gohier/National Audubon Society Collection/PR; *bottom,* Andrew Laurie. **(176)** PITCH/Francois Gohier. **(177)** Tetsuo Gyoda/ORION PRESS. **(178)** Jen & Des Bartlett. **(179)** *left,* Stanley Breeden; *right,* Pauline R. McCann/BRUCE COLEMAN INC. **(180-181)** © Reinhard Kunkel. **(181)** *top,* © Reinhard Kunkel; *bottom,* Patricia D. Moehlman. **(182)** *top left,* Richard W. Brown; *top right,* Hans Reinhard/BRUCE COLEMAN INC.; *bottom left,* © Stephen Green-Armytage 1981; *bottom right,* James L. Stanfield/N.G.S. Photographer. **(183)** *top,* © Stephen Green-Armytage 1981; *bottom,* Momatiuk/Eastcott/Woodfin Camp Inc.

(184) Kim Taylor/BRUCE COLEMAN INC. **(185)** *top,* PITCH/J. L. Blanchet; *bottom,* Rene-Pierre Bille. **(186-187)** Ben Cropp. **(188)** *top,* Brian J. Coates/BRUCE COLEMAN LTD.; *bottom,* Jen & Des Bartlett. **(189)** *top,* PITCH/Jean-Paul Ferrero; *bottom,* Jen & Des Bartlett. **(190)** © Peter Johnson. **(190-191)** George W. Frame. **(192)** *top,* © Patricia D. Moehlman; *bottom,* © Reinhard Kunkel. **(193)** M. P. Kahl. **(194)** © Reinhard Kunkel. **(194-195)** Bob Campbell. **(195)** *top,* © Peter Johnson; *bottom,* F. J. Weyerhaeuser. **(196)** Stanley Breeden. **(197)** Dieter & Mary Plage/BRUCE COLEMAN LTD. **(198-199)** Steven C. Wilson/ENTHEOS.

(200) Steven Fuller. **(201)** *top,* Gary R. Zahm; *bottom,* Charles G. Summers, Jr. **(202-203)** *sequence,* B. J. Rose. **(203)** ANIMALS ANIMALS/Michael & Barbara Reed. **(204)** Peter Ward/BRUCE COLEMAN INC. **(205)** Zoological Society of San Diego. **(206-207)** Farrell Grehan. **(207)** *top,* Wolfgang Bayer; *bottom,* Lynn L. Rogers. **(208)** *top,* Glenn D. Chambers; *bottom left,* Jen & Des Bartlett; *bottom right,* ANIMALS ANIMALS/John C. Stevenson. **(209)** *left,* P. Morris/ARDEA LONDON; *top right,* © Reinhard Kunkel; *bottom right,* Steve Maslowski. **(210-211)** Robert A. Garrott; *inset,* Douglas H. Chadwick. **(212)** © 1978 Patricia D. Moehlman. **(212-213)** M. P. Kahl. **(214)** Gerald Cubitt. **(215)** *top,* PITCH/P. Montoya; *bottom,* PITCH/F. Charmoy. **(217)** George Holton /National Audubon Society Collection/PR. **(218)** Edward S. Ross. **(219)** *top,* Carol Hughes/BRUCE COLEMAN LTD.; *bottom,* Moira & Rod Borland/BRUCE COLEMAN INC. **(220)** PITCH/P. Montoya. **(221)** Peter Davey/BRUCE COLEMAN LTD. **(222)** Tom McHugh/National Audubon Society Collection/PR. **(223)** Teleki-Baldwin. **(224-225)** Tom McHugh/National Audubon Society Collection/PR. **(226-227)** © Peter Ward/BRUCE COLEMAN LTD. **(228)** Ian Beames/ARDEA LONDON. **(229)** *top,* Stewart Cassidy; *bottom,* © 1978 Patricia D. Moehlman. **(230)** *top,* M. P. Kahl/National Audubon Society Collection/PR; *bottom,* Clem Haagner/BRUCE COLEMAN INC. **(231)** Gerald Cubitt. **(232)** Douglas H. Chadwick. **(233)** Tom & Pat Leeson. **(234)** Keith Gunnar; *inset,* Karen B. Reeves. **(234-235)** Tom & Pat Leeson. **(236)** Douglas H. Chadwick. **(237)** *left,* Kenneth W. Fink/BRUCE COLEMAN INC.; *right,* © Robert L. Dunne/BRUCE COLEMAN INC. **(238-239)** George B. Schaller. **(239)** *top,* George B. Schaller; *center,* Kenneth W. Fink/ARDEA LONDON; *bottom,* Javier Andrada. **(240)** Craig R. Sholley. **(241)** *top,* Dian Fossey; *bottom,* David P. Watts. **(242-243)** David P. Watts; *inset,* Craig R. Sholley. **(244)** *top,* David P. Watts; *bottom,* Craig R. Sholley. **(245)** Craig R. Sholley. **(246)** *top,* PITCH/Francois Gohier; *bottom,* Tony Morrison. **(247)** Jean-Paul Ferrero/ARDEA LONDON. **(248)** Jen & Des Bartlett. **(249)** S. C. Bisserot.

(250) Jane Burton/BRUCE COLEMAN LTD. **(251)** Thomas D. Mangelsen. **(252-253)** Rene-Pierre Bille. **(253)** *left,* Rollie Ostermick; *center,* Stephen J. Krasemann/DRK Photo; *right,* Franz J. Camenzind. **(254)** *left,* Clem Haagner; *top right,* Joanna Van Gruisen/ARDEA LONDON; *center,* M.P.L. Fogden; *bottom left,* Derek Middleton/BRUCE COLEMAN LTD.; *bottom right,* PITCH/Cordier. **(255)** *top,* David R. Gray; *bottom,* Larry R. Ditto. **(256)** *top,* Hans Reinhard/BRUCE COLEMAN LTD.; *bottom,* S. C. Bisserot/BRUCE COLEMAN LTD. **(257)** *top left and right,* Peter Ward/BRUCE COLEMAN INC.; *bottom left,* PITCH/J. Delacour. **(258)** Gerald Cubitt. **(259)** © Peter Johnson. **(260)** © Peter Johnson. **(260-261)** Hervy/Jacana/The Image Bank. **(262-263)** ARDEA LONDON. **(262)** Alan Root. **(263)** Philip Coffey. **(264)** PITCH/Billes. **(265)** Ivan Polunin/NHPA. **(266-267)** T. W. Ransom. **(266)** *top,* Gerald Cubitt; *bottom,* Bildarchio Paysan Stuttgart. **(267)** Gerald Cubitt. **(268)** *top,* Webbphotos/Ted McDonough; *bottom,* Frank W. Lane. **(268-269)** Victor Englebert. **(270-271)** © Jonathan T. Wright/BRUCE COLEMAN INC. **(272-273)** © Stephen Green-Armytage 1981. **(274)** *left,* H. W. Silvester/RAPHO; *right,* James L. Stanfield/N.G.S. Photographer. **(274-275)** © Elisabeth Weiland/National Audubon Society Collection/PR. **(275)** George B. Schaller. **(276)** *top,* Sally Anne Thompson/Animal Photography Ltd.; *bottom,* Thase Daniel. **(277)** Tom McHugh/National Audubon Society Collection/PR. **(278-279)** *top,* Carol Hughes/BRUCE COLEMAN LTD.; *bottom,* George W. Frame. **(279)** *top,* Robert Caputo; *bottom,* David Bygott. **(280)** M. P. Kahl. **(280-281)** David Bygott. **(282)** *top,* Clem Haagner; *bottom,* Philip Coffey. **(283)** Hendrik N. Hoeck. **(284-285)** Peter Davey/BRUCE COLEMAN LTD. **(284)** *bottom left,* © Gail Rubin; *bottom right,* Bob Campbell. **(285)** © Gail Rubin. **(287)** © Jean-Paul Ferrero. **(288)** *top,* © Jean-Paul Ferrero; *bottom,* Rene-Pierre Bille. **(289)** © 1977 Gail Rubin. **(290-291)** © 1977 Gail Rubin. **(292)** M. P. Kahl. **(293)** Clem Haagner/BRUCE COLEMAN INC. **(294)** M. P. Kahl. **(294-295)** © Reinhard Kunkel. **(295)** Wolfgang Bayer. **(296)** © 1978 Patricia D. Moehlman. **(297)** © Reinhard Kunkel. **(298-299)** © 1978 Patricia D. Moehlman. **(300-301)** Loren McIntyre. **(302)** Loren McIntyre. **(303)** Mark Rosenthal-Kimberly Lile. **(304)** Irven DeVore/Anthro-Photo.